Great Modern Inventions

Other titles in
Chambers Compact Reference

To be published in 1993

Great Modern Inventions

Gerald Messadié

Chambers

EDINBURGH   NEW YORK   TORONTO

Published 1991 by W & R Chambers Ltd,
43–45 Annandale Street, Edinburgh EH7 4AZ
95 Madison Avenue, New York N.Y. 10016

First published in France as *Les grandes inventions du monde
moderne*

© Bordas, Paris, 1988

© English text edition W & R Chambers 1991

Library of Congress Cataloging-in-Publication Data applied for

ISBN 0550 17001 4

Cover design Blue Peach Design Consultants Ltd
Typeset by Hewer Text Composition Services, Edinburgh
Printed in England by Clays Ltd, St Ives, plc

Acknowledgements

Translated from the French by Melanie Hanbury

Adapted for the English edition by Bill Houston
Min Lee
Penny Smith

Entries provided for the English edition by Min Lee:

Aspartame	Fatless meat
Breakfast cereals	Food processor
Carpet sweeper	Lego
Dishwasher	Microwave oven
Electric hand drill	Non-stick pan
Electric heater	Velcro®
Electric mixer	Videophone
Escalator	

Terence Lee:
Szilard's engine

Jack Osborne:
Collotype
Linotype
Phototypesetting

Chambers Compact Reference Series Editor   Min Lee

Illustration credits

Page

33 © DITE
42 © Archives Photeb
48 top left: © Musée national des
Techniques Collection, CNAM,
Paris/Archives Photeb
48 © DITE
55 top: © Roger-Viollet/Archives
Photeb
55 bottom: © Thomson-CSF/Archives
Photeb
58 © PPP-IPS/Archives Photeb
60 © DITE
62 © CDHT-CNAM Collection, Paris/
Photeb
63 top: © DITE
63 bottom: © DITE
67 © SIC Postes Télécom Espace
75 © Coll. PPP-IPS/NASA/Archives
Photeb
81 © Scriptel Holding Inc./Photeb

Page

85 © Jean Pottier/REA
89 © Honda
149 top: © Institut Géographique
National
149 bottom: © Institut Géographique
National
157 © IBM
160 © JEOL
164 © DITE
167 © DITE
198 © DITE
201 © Musée de la Marine, Paris/Photeb
205 © CDHT-CNAM, Paris/Photeb
210 © Focus/T
211 © Citroën
214 © Northrop
217 © PPP-IPS/Archives Photeb
219 © DITE
223 © Viollet/Archives Photeb Collection
224 © AFP

Front cover (light bulb) Garry Gay © The Image Bank
Spine (automatic telephone) Lausat © Explorer

Contents

This volume is the sequel to *Great Inventions through History*, also in this series.

Industry and Industrial Technology

Instruments for Measuring and Observation

Medicine and Biology

Transport

Warfare and Military Technology

Editor's note
In each article in this book, the reader will find words highlighted in **bold type**. These indicate the principal stages (references, objects, circumstances and aims . . .) which have contributed to the invention.

Introduction

Dividing inventions into two historical categories and placing each in a separate work, the one comprising inventions before 1850, the other, those which came after that date, is a somewhat arbitrary act. However, the year 1900 appears from our modern vantage point as the time when the technosphere entered, first in rivalry, and then in to conflict, with the biosphere.

During the preceding thousands of years things were invented and manufactured; but the fires in potters' workshops, for example, never threatened to obscure the sky, nor was the apothecaries' research likely to upset population balance, or the economy. In the 19th century, however, London became one of the most polluted cities in the world. The famous 'fog' along the banks of the River Thames actually consisted of a mixture of fog and industrial fumes in a stagnant atmosphere, which these days is called 'smog'; effective measures against it have been taken during the second half of the 20th century, so that the *fog* that was so dear to readers of horror stories and crime thrillers has effectively disappeared. In Paris at the same time, the River Seine was an open-air sewer, the stench from which suffocated the city dwellers: it was caused by the disposal of excrement and waste water directly into the river, a situation made worse by the increasing population. This increase was a result of an increase in human life expectancy, which was partly due to medical advances. The Seine did not begin to be odour-free again until around 1920 and even if it is only a fool who will eat a gudgeon for which he has fished (in fact it is particularly in the River Marne that this fish can still be found), from this time onwards it was possible to have a picnic at Charenton or at Robinson on the banks of the Seine without fainting.

As the 20th century advanced, the conflict between the technosphere and the biosphere became more pronounced. During the 1950s the publication of Rachel Carson's *Silent Spring*, a book which describes the horrifying damage done by DDT to the natural environment, aroused great alarm in the Western world. DDT had seemed indispensable in the battle against so-called harmful insects. Hardly 15 years separated these first alarms from those which were to be aroused by the building of nuclear power stations. In 1973 after a serious accident had happened at Windscale Power Station in Cumberland, UK, panic spread throughout the country and into Europe: what if this type of accident should happen again? It did happen again, in fact, and not a month passes without some problem, albeit a minor one, arising in a nuclear power station. The governments and the scientists who served them, started to lie. The bad name of Windscale was removed by its being renamed as Sellafield; its original name is even disappearing from reference books. Accidents become more and more serious: after that at Three Mile Island in the United States, came the disaster at Chernobyl in the USSR in 1986.

The dubious consequences of progress

Since the 1970s people have been worrying about the effects of excessive processing of food products, and consumer action has forced governments to ban certain non-essential additives. Similarly, it was found that the chlorofluorocarbons, which were used in the manufacture of a particular type of aerosol, might have dramatic repercussions on the future of the whole of humanity: their use was helping to cause the destruction of the layer of ozone in the atmosphere which acts as a filter for ultraviolet rays. One result of this could be an increased risk of skin cancer for susceptible individuals. Industrial pollutants are strongly suspected of being the cause of acid rain which, in turn, causes massive deforestation in far away lands. And the harm caused by the lead content in car exhaust fumes is leading to nearly all the governments of industrial countries encouraging the use of unleaded petrol.

For the first time ever, mankind is realizing that the technosphere which he himself has created may endanger the biosphere in which he lives. Catastrophic river pollution, the accident at Bhopal in India, and the damage caused to flora and fauna by insecticides have reinforced the ecological concern so that there is good reason to hope that it will influence government policies in the future.

The effects of the technosphere are not all negative; far from it. Since the end of the 19th century the spread of vaccines and a continually improving knowledge of hygiene have increased life expectancy from birth. Similarly, the obligatory measures of asepsis and antisepsis in maternity hospitals have resulted in a considerable reduction in infant mortality. In the 18th century a child only had a one-in-five chance of reaching the age of one, and, having reached it, had an average life span of 45 years, but by the beginning of the 20th century it had one chance in three and its life expectancy was increased by ten years. Three quarters of a century later in the developed countries it was to increase again by 20 years. Indeed, in the developed countries, a person's life expectancy will have increased by 30 years since the beginning of the 20th century. We owe the 20 years gained between 1940 and 1980 (approximately, for figures vary according to the country) to antibiotics, better disease detection, progress in surgery, pharmacology, food technology and prevention of disease, as well as to education.

The consequences of this are immense, firstly from a demographic point of view. Between the beginning of the century and the start of its final quarter, an incredible difference can be seen in the demographic evolution of the developed countries compared to those which are described as 'emerging'. In 1940 France had a population of 43 million, and Egypt had one of 16 million. In 1985 France had 53 million inhabitants, and Egypt, 45 million. It is as though the scientific and technical development and consequently the growth of life expectancy were encouraging the people of the developed countries to control their demographic growth, whereas the countries without these advantages were trying to compensate for their handicap by producing more people. In Egypt in 1980 life expectancy was not even 50 years — 25 years behind Europe! The development of the technosphere has thus altered the human balance in the biosphere, and also the economic and political situation. This was particularly noticeable during the 1970s when South East Asia began to manufacture electronic goods at such low prices that the developed countries could not compete. By 1985 the United States had relinquished nearly all of the electronic industry to Asia and now only makes one type of television, and a single radio set; the situation in optics is nearly the same. The immediate result is that the research centres have moved towards the new economic centres, and Japan occupies a high position in numerous scientific and technological areas, such as the making of the so-called Fifth-Generation computers.

Even there the economic consequences of the development of the technosphere are tackled. Beyond the fact that the countries which were traditionally the technological innovators are now less important, a second consequence must be noted: the significant rise in the quality of life. Automation has allowed the reduction, sometimes in proportions which range from 1 to 10, of the cost price and therefore the selling price of consumer goods. At the same time, automation has created unemployment and has therefore subsequently slowed down its own expansion. The unemployed are unable to be consumers, consumption is slowed down and, to a certain extent, so is technological innovation.

The age of permanent recycling

If this landscape is compared with that of previous centuries one can see not only breathtaking differences but also a caesura which once more justifies establishing a border between the 20th century and the preceding centuries. No invention in the past had created unemployment nor threatened the environment. Sometimes the situation created by automation in the

20th century is compared to that which resulted in the 17th century following the introduction of the new weaving looms. In actual fact, the two are very different, for the looms were a threat to the skilled handiwork of the traditional weaver, to the advantage of the unskilled labourers. Exactly the opposite has happened in the 20th century: technology favours the most skilled workers to the detriment of those who are less skilled.

In the 20th century, the working knowledge which is passed on to graduates, has to be updated more and more frequently. At the end of the 1980s a diploma was considered to have an invisible expiry date of a minimum of three years and a maximum of seven years following the year it was awarded. In electronics for example, the arrival of optical fibres in computer design and the promise of superconductive chips meant that a great many electronics engineers were obliged to acquire new knowledge in their field.

In many leading domains invention has managed to be considered only as the fruit of permanent research. People are no longer 'scientists' but 'researchers', and they remain so from beginning to end of their career. This has not always been the case. Even in the 19th century when Edison had invented his electric light bulb (admittedly at the cost of thousands of experiments), he would have been at liberty to 'rest on his laurels', if he wished, for several years. He could hope for improvements in the vacuum pumps, filaments, glass, in other words, evolutions, but not permanent revolutions. In 1987, on the other hand, a press conference on the new superconductors which was held at Kyoto became a riot: when Dr Ihara announced that he had observed the phenomenon of superconductivity at the unthinkable temperature of 65K, international pandemonium ensued. The existence of materials capable of conducting electricity without resistance, not at very low and costly temperatures, but at ordinary temperatures would lead to the design of new generations of batteries, new means of electric transport, electric cars which could keep moving for 5000 km, magnetic levitation trains which could travel at 400 km/h, computers capable of carrying out operations in real time, and so on.

In another domain, that of biology, it is worth noting that at the beginning of the 20th century a vaccine was made by inoculating with a bacterium which had previously been killed or weakened in certain aseptic conditions. In 1987 no less than 30 different methods were available to make a vaccine. Each one of these methods could entail virology and a whole section of molecular biology proceeding along entirely new routes. The technosphere is thus becoming more and more tense.

Technology must be adapted to geography

The differences between 20th century inventions and those of the preceding thousands of years certainly do not end there. A profound change must be added, which this time occurred in people's mentalities.

From the Industrial Revolution until around 1960, scientists, technicians, inventors and the public all entertained the conviction, which was inherited from industrial 'St. Simonism', that science and technology were always at the service of man and of society. Inventions, all inventions, could therefore be beneficial to humanity, including the invention of methods of nuclear fission, which could provide cheap energy. Then, slowly, from around 1960, it turned out that the technological packages which were exported at great expense to the developing countries, more commonly referred to then as the Third World, were not producing the expected benefits. Airports, dams, electrification, telemetric networks, etc. were added on to social structures and ancient cultures, sometimes disrupting, or even destroying them, but only rarely benefiting the whole population. Brazil, for instance, is a country which is equipped with up-to-date technology and exports armaments, aeroplanes, motor cars and computers, but its wealth is concentrated among little more than 20% of the population.

Consequently, firstly in the agricultural domain and then in the whole of the technological sphere, experts expressed the opinion that it was not useful and it was perhaps harmful to impose the technological structures of industrial countries onto the Third World. In agriculture for example, the irrigation techniques using the antiquated *chadouf* could be much more effectively adapted to the developing countries, because they require next to no investment and hardly any costly labour compared to methods based on the use of mechanical pumps. The usefulness of these pumps in industrial countries can hardly be contested, but in tropical countries heat pumps are considered to be preferable.

The example of the high dam in Aswan in Egypt also makes one wonder about technology. It was built in 1963 to increase the country's production of electricity and to help the creation of new farming land by stopping the loss of water which accompanied the Nile floods. The two aims were achieved, but it seemed that the mass of water induced infiltrations of soft water into the soil which dissolved the underlying layers of salt and caused salt water flushing which was actually harmful to the land under cultivation. In other words, an appreciable part of that which was gained in workable surface time was lost in agricultural yield. This was one more time when it became obvious that an an invention has disadvantages as well as advantages.

Secondly, certain analysts extended their works to the industrial countries themselves, the creators and exporters of technology, and also of inventions. The United States occupies the first place among these countries. However, some American onlookers have highlighted the fact that about 40% of the American population hardly benefit at all from advanced American technology. One of the fundamental reasons is that research for industrial purposes, which can be simply referred to as research & development, absorbs a huge part of the government budget — half of it in the United States and the United Kingdom, a little less in France. And the more progress there is in technology, the more the budget for R & D grows.

That amounts to saying that, for the first time in history, technology, the mother of invention, is considered as a domain which is not automatically going to contribute to the well-being of society and which can even be dangerous to it, when it produced massive means of destruction. In the majority of countries which have a large research budget, military purposes are dominant.

This state of affairs has captured the attention of numerous specialists, scientific historians, philosophers and sociologists.

The age of collective inventions

Finally, a great change in the domain of inventions should be noted: they are becoming more and more collective. In the past, unless it was withdrawn, the inventor's name was always connected to the invention: for example, James Watt and the steam engine, Stephenson and the locomotive, Marconi and wireless telegraphy. The financial investment necessary to invent something was actually within the reach of a (reasonably wealthy) individual. This was the case until the first decades of the 20th century; modern rocket technology began with work by Robert Goddard, whose experiments were funded in quite a modest way by the Smithsonian Institution.

As technology progresses however, inventions require increasingly significant and expensive resources. Leeuwenhoek invented the optical microscope without becoming bankrupt; but Knoll and Ruska, who invented the electron microscope in 1932, could not do so without the use of some very high-level equipment which they could not have bought for themselves. Tradition makes it desirable that in certain cases the name of an individual who makes an invention using a large firm's materials should be associated with the invention which automatically becomes the property of the firm. But for a variety of juridical and financial reasons, this is less and less often the case. Inventions tend to be presented under the name of the firm where they have been made. This is sometimes because there are several inventors who have worked in a team, and sometimes because, even if the invention is due to one single inventor, it is the result of research financed by the firm which considers itself to be its legitimate owner.

It is likely, therefore, that in the 21st century the majority of inventions will be the property of those large firms that have the means to carry out considerable research. Thus the invention will pass into the control of large multinational firms or state institutions.

Invisible inventions: automation and miniaturization

Great Inventions Through History, also in this series, presents a particular subject for consideration: the great antiquity of some inventions which for a long time have been considered to be 'modern', such as the universal joint or the steam machine. This book offers another: that of the growing technical complexity of inventions which renders them almost invisible.

Thus, with the exception of the replacement of propeller motors by turbojets, there seems to be no essential difference between a Potez dating from one of the commercial airlines of 1939 and a Boeing 747 or an Airbus of the same airline in 1990. However the second two differ considerably from the former: the automatic systems have been multiplied and include an automatic balancing system which makes the aeroplane more stable. Automatic regulation was not 'invented' in the 20th century; in fact it has existed since the float regulator of the School of Alexandria. But it is during the 20th century that its use has become widely spread.

The same applies to miniaturization, which was not invented by anyone. Very strictly speaking, the Americans, Bardeen, Brattain and Shockley, who invented the transistor could be designated as its putative fathers, but these three physicists stopped at the invention of the transistor, which succeeded the old triode. The fact remains that miniaturiation would affect a huge part of 20th century industrial production, particularly in the domain of electronics.

It is impossible to find an inventor of the numerically controlled machine tool which works from the instructions of a punched paper tape (it began to be used in the 1940s and has developed into CNC or computer numerically controlled machine tool). Perhaps its inventor is none other than the visionary mechanic of the 18th century, Jacques de Vaucanson, who was responsible for the first automatic weaving loom, which worked on perforated cards. In this way, a product which seems quite recent is often no more than the latest version of a very old invention.

An important feature of 20th century technology is that it has become almost uncontrollable. It follows its own autonomous evolution, comparable in this to the functioning of DNA, which seems to be 'indifferent' to the fate of the cells, the production of which it controls (to the point that some great biologists have also referred to 'the selfish gene').

An uncontrollable and unpredictable evolution

This final, undeniably troubling feature is the unpredictability of science. Even though it is true that thousands of scientists all over the world are endeavouring to find solutions to certain problems, such as cancer, it is often still the case with inventions that it is fate which calls the tune. Therefore, in compiling a list of all the human genes and the relationship between several of their anomalies and certain illnesses, it is found that there are some genes which are linked to cancer of the uterus, the breast, or colon, etc. The remedy is certainly not to be found where it is being sought; it would undoubtedly lie in the correction of the defective gene. It was also by chance that Bernard Raveau discovered a material which was superconductive at temperatures well above the usual temperatures for superconductivity. Since one can hardly invent except from that which has already been discovered, it follows that inventions are off-shoots of discoveries and largely also of chance.

Not only are they born of chance, but even then they seem to take on an impetus of their own and become established, whatever the real needs of the community. The most famous example is that of the motor car, the usefulness of which is becoming geometrically inverse to the numbers on the road, but which continue to feature in public opinion as a necessity of everyday life. Another, less dramatic example is that of the short- and medium-haul aeroplane. In the United States, for example, airlines have created such congestion that sometimes, taking into account the delays and the transport time from the airport to the city centres, they take three times the theoretical time of the journey to get from one town to the other. Meanwhile a train, which would carry at least five times more passengers on each journey, would be on time and would take the travellers from city centre to city centre with less risks. But the aeroplane has the image of speed and so is preferred.

The history of 20th century inventions includes a lesson in philosophy, whilst that of the preceding centuries includes one in psychology. In considering the past, one can only marvel at the precocity, albeit unwitting, of technical inventiveness, and also at the strange slowness with which some of the major inventions have come to light. In recent decades it is surprising what modest beginnings some inventions have had, and the tremendous (even excessive) influence which they have exercised over time.

A new branch of study has developed in this century reflecting the destiny of the human race: the philosophy of science.

Agriculture and the Food Industry

It might seem surprising that from the second half of the 19th century to the present day, little progress has been made in the way of inventions in agriculture and the food industry. In fact, setting aside the mechanization of agriculture and the canning process, which came earlier, it can be said that these two areas of human activity have not been changed by any major invention at all. The huge differences between 19th and 20th century agricultural methods lie in the improvement of technique and variety. Even the creation of brand new species, which is the peak of agricultural progress in the 20th century, is nothing but a spin-off of genetic engineering; nor has it made any significant difference. Some experts are actually alarmed at the fact that the cultivation of certain varieties is tending to impoverish the genetic environment of other vegetable and animal species.

In the food industry, the advent of industrial irradiation at the end of the 20th century is just the development of an idea that goes back to 1896! Even if human tastes in food change considerably, the basic foods eaten remain essentially the same, and the ways of preserving and preparing them have hardly altered since the 19th century. Many people do not eat pork cut to order any more; often it is packaged ready-sliced — but it is still pork. Thanks to increasingly strict sanitary controls, the quality of the meat is undoubtedly improved. It is also standardized, to a large extent, with many butchers and supermarkets supplying meat from similar sources.

Aspartame
Searle Laboratory, 1965

'Artificial' sweeteners, or sweeteners which do not contain sugar, are of great interest to the food industry for the huge 'diet' market. Aspartame (marketed under the brand name Nutrasweet®) has 200 times the sweetening power of sugar. Prepared from two amino acids, phenylanyline and aspartic acid, it was made by scientists at the Searle Laboratory in the USA in 1965. It does not have any bitter aftertaste, so is ideal for use in processed foods. It was licensed for use in the USA in 1982, and is also permitted in the UK, Germany and Switzerland. It has no known side effects, except for sufferers from phenylketonuria (PHK).

Cafeteria
Harvey, 1876

The invention of the cafeteria, an establishment in which a set number of dishes is served and where the service is partially assured by the customer himself, is undoubtedly one which has greatly altered the food industry and customs in general all the world over. Previously, the only eating places were restaurants which had relatively varied menus and where the service was totally entrusted to qualified staff.

The invention is essentially American and it was in the United States that cafeterias first appeared and were subsequently called **automats**. The first commonly known ones seem to be those opened in 1876 by Fred Harvey, commissioner of the Chicago, Burlington and Quincey railway line; they were situated at the stations of the then well-known Atchison, Topeka and Santa Fe line. Travellers could refresh themselves quickly and adequately at little cost. These were, in a way, modified cafés, whence ultimately their name derived. It is possible perhaps that Harvey was inspired by the **refreshment trolleys** drawn by animals, which first appeared in Providence, Rhode Island, in 1872, and which sold slices of cold chicken, hard-boiled eggs and sandwiches to pedestrians.

One social factor in particular contributed greatly to the spread of cafeterias: no alcohol was served, the food was healthy and the places impeccably clean, but, above all, the standards of dress and· correct behaviour were faultless; these were establishments where single women could go without fear of being troubled or seeming to be on the search for romance. This was not the case in saloon bars, where food was thought of as a mere accompaniment to alcohol and where women were not safe, nor in all other restaurants, where women alone tended to attract the attention of masculine customers. They were, therefore, 'bourgeois' establishments and in fact the new American bourgeoisie, small, average and desperate for respectability, assured them a numerous and steady clientele. The Church Temperance Society, a Protestant league for temperance 'opened eight of them as alternatives to patronizing the saloon bars', wrote the American historians, Jane and Michael Stern.

Another social factor also furthered the expansion of cafeterias; this was the accelerating rhythm of urban life which went with the increasing number of people working in offices. The 'white-collar workers', the common nickname for office employees in the United States, who had neither the time to go home for lunch, nor the financial means to frequent restaurants, and who wanted to remain sober, encouraged the establishment of more and more cafeterias in the business areas. The form varied according to the establishment and the town: in the Mid-West, the customers helped themselves at a counter, paid at a till and then sat down at an empty table; in New York in 1890, however, the Exchange Buffet introduced the idea of eating at counters whilst remaining standing, which ruled out the need for individual tables to be installed and allowed the use of the available area in a much more profitable way. It is a system that is now a great success in the majority of Western countries.

The word 'cafeteria' itself was however only coined in 1902 by the American, John Kruger, who had followed a Scandinavian model for the meals and the layout of his establishment. Kruger had first thought of calling it 'smörgåsbord', but his compatriots found the word difficult to pronounce, so he replaced it with the Spanish *cafetería* which means, quite simply, 'café'.

The cafeterias quite clearly changed people's eating habits by introducing the concept of a meal consisting of a meat dish and vegetables or salad, with no alcohol. This intensified the **standardization** of food which had already been imposed by the food preserving industry. Different meats were prepared in a similar way and the sizes and preparation methods of vegetables were more or less standardized too.

The cafeteria system only began to take root in Europe in the 1960s, especially as a result of growing dietary concerns and of young people beginning to have a more independent social life.

Three new factors meant that at the end of the 1970s the cafeteria system was perhaps becoming out-dated. The first was the rise in the standard of living, at least in the United States, which resulted in the excessively 'functional' character of this type of establishment losing its appeal in favour of proper restaurants. The second was the trend of the townspeople to move house to the suburbs which, paradoxically, decreased business at the cafeterias; in fact, the shortening of the working day, with the lunch break reduced to half an hour, increased the habit of eating sandwiches or snacks at the work place. The third was the introduction of packaged ready-cooked meals and the new variety of preserved food, which meant that those clients who went to cafeterias because they did not want to eat out of a tin can at home tended to change their ways.

By the end of the 1980s, part of the cafeteria market had given way to the system of food being sold to be taken away, that is, **fast food**.

Clementine

Clément, 1900

Most probably a hybrid of the bitter orange (from which neroli oil, used in perfume, is obtained) and the mandarin orange tree, the clementine tree was produced in 1900 after much research by Clément, a priest from Oran. An early producer, its fruit, the clementine, is akin to the mandarin and has a delicate flavour; it has spread over nearly all western Mediterranean countries.

Fatless meat

Chapman, 1988

The purpose of this product is to provide a low-calorie, low cholesterol alternative to natural meat. An Australian butcher, Dallas Chapman, carried out research for 15 years before creating a form of meat which looks like sausage meat, but in which the fat content is reduced by 96%, the cholesterol by 30% and the calories by 85%.

Food additives

Anon., c.1860

The phrase 'food additives' should be understood both in the broad and the narrow sense. Broadly speaking, with some additives having been used since early history, it is not a modern invention. In this sense salt is a preserving agent, and spices, herbs and seasonings are flavour enhancers. It was necessity that prompted the use of these two types of agent, which is not always the case when additives are used nowadays. Modern additives are often synthetic **colourings, preserving agents** (fungicides, bactericides, preservatives, antioxidants), **texture-improving agents** (moistening agents, emulsifiers, stabilizers, thickeners), **ripening agents**, and **flavouring agents**, all of which are generally of artificial origin.

It was only when **industrial food production** first began, during the second half of the 19th century, that the first of these types of food additives appeared. These were mainly industrial **colourings** used in confectionery (sweets made with chromic or lead oxide) added to recreate the symbolic factors associated with the taste of the food: the red of lead for sweets meant to be strawberry flavoured, blues and purples for bilberry, etc. In 1877, garden peas were made green with copper salts! This method of colouring food even resulted in a red cheese tinged with mercury sulphate, which was no less toxic than the other colourings, for it too consisted of a heavy metal. During the 1920s the use of **butter-yellow** (phenyldimethylaminobenzene) became widespread. It is now banned because it was discovered that this colouring, which was used to give butter that was naturally pale in colour the appearance of 'golden' butter, could cause the growth of tumours in experimental animals.

The use of colourings continued nonetheless in most countries and particularly in industrial countries, owing to the poor development of toxicology. **Copper sulphate** was formerly used in food although it is dangerous to man and animals. **Agene** (nitrogen trichloride) was used as an improver in most of the functions to make bread in the United Kingdom, until it was established as a cause of damage to the nervous system in dogs.

Since the 1960s, following pressure from consumer associations, studies in toxicology have increased and resulted in the direct banning of some colourings and the restricted use of others. Several countries, such as Greece and Norway, prohibit the use of all food colourings; others ban only specific ones. International legislation is a long way from being standardized. The yellow colouring, **tartrazine** (E 102), is still allowed in many countries including the UK although it can cause serious allergic reactions.

The use of fungicides is restricted and carefully controlled, but the preservative sorbic acid (E 200) and its salts (E 201 to 203) are found in many industrially produced foods because they inhibit growth of yeasts and moulds. Benzoic acid (E 210), and its derivatives or benzoates (E 211–218) are found in products where discolouration is to be avoided, such as jams, fruit juices and pickles. It was the joint action of industrial food firms and many fundamental and applied experiments which brought about their use in the first place. In France the use of sorbic acid in baking has been banned for twenty years since it was proved that it was likely to react with the preserving nitrites to produce a mutagen.

The first of the **anti-oxidants** to be used in the food industry was vitamin C or **ascorbic acid** (E 300), used first of all in the USA in the 1940s (synthesis only became possible in 1932). It was to come into general use worldwide after World War II, together with some of its derivatives, such as calcium ascorbate (E 303) and ascorbyl palmitate (E 304) which are esters. Towards the end of the 1960s, vitamin E (E 306), extracted from soya or wheatgerm, began to be used, as well as its derivatives, the other tocopherols (E 307, E 308 and E 309). Some synthetic antioxidants such as **E 310**, which was patented in 1942, are now restricted in their use: **E 310**, propyl gallate is not permitted in foods intended for babies and young children. Related chemicals (E 311, E 312, E 320, E 321, E 336) are authorized for use in quantities which are constantly changed. Their use is contested because they seem to have allergenic effects. The additive **E 311** in particular may cause gastric irritation, and problems for people who suffer from asthma, or are sensitive to aspirin.

Moistening agents, thickeners, emulsifiers, etc. constitute the group of **texture-improving agents**, which subdivides into four large groups: anti-agglomerates, thickeners, stabilizers and gelling agents, and the actual emulsifiers. There is controversy about thickeners and gelling agents, some of which are suspected of being allergenic. Several of these are extracts of natural products such as carob flour, guar gum and **karaya gum** (banned in France and some other countries of the EC, but not in the UK); their natural origin does not mean that they are devoid of side-effects, whether positive or negative. The use of the first texture-improving agents is very old — flour has been used to thicken sauces since ancient times, and proof of the use of jellies, which were formerly made with egg white, then with bone stock, has been discovered in medieval cooking recipes. **Lecithin** (E 322), used as an emulsifier, can be extracted from egg yolk; most of that which is used in the food industry is extracted from soya by means of chemical solvents. It was at the end of the 19th century that texture-improving agents were improved and perfected: monoglyceride and diglycerides of fatty acids in foods (E 471), acetic acid esters and lactic acid esters (E 472), esters of propyleneglycol (E 477), sucroesters and sucroglycerides (E 473 and E 474). **Polyphosphates** (E 450), which are seldom quoted in the list of contents declared on packages, serve to retain water during cooking and prevent meats from becoming stringy-looking. They are used a great deal in the delicatessen trade. It has been suggested that polyphosphates could cause digestive problems by blocking certain enzymes.

Ripening agents are chemical substances added to fruits and vegetables whilst they are growing in order that they grow uniformly; they often have multiple effects, serving also to repel insects, prevent moulds, which in some cases puts them in the class of fungicides, and to ensure that the fruit travels well. The procedure is complex and varies from country to country. In 1989, one of these agents, **diaminozide**, which was on 5% of the red apple harvest in the United States, was taken off the market because it had been found to produce cancer in experimental animals.

Flavour enhancers are meant to 'bring out' the taste in foods. They are either acids (acetic acid E 260, acetates E 261–263, lactic acid E 270 and lactates E 325–327, citric acid E 330 and citrates E 331–333, tartaric acid E 334 and tartrates E 335–337, phosphoric acid E 338 and orthophosphates E 339–341) or sugars (sorbitol E 420, glycerol E 422), or are meant to accentuate the taste as do **glutamates**. These are obtained commercially from the fermentation of sugar beet or wheat gluten; monosodium glutamate occurs naturally in a seaweed used in Japan.

At the end of 1988, neurologists analysed the role of glutamates. These are substances which excite the nerve endings and send amplified messages to the brain, which lead to the impression of having a more intense experience of the taste of the food; they are also called taste exhausters. In some people they may give rise to a phenomenon known as **'Chinese restaurant syndrome'** which manifests itself in headaches, hot flushes, nausea, disturbance of vision and dizzy spells.

Several countries allow additives in certain foods, such as bread, confectionery, milk and margarine, in the form of **vitamins** or iron. Medical authorities have greatly reduced these additions. All vitamins, with the exception of vitamin C and perhaps vitamin E, can potentially cause hypervitaminosis and **iron** has been clearly contraindicated in certain metabolic problems, as well as cancers.

Food additives proceed alongside other technological and economic modifications. Their purpose was to allow food to be kept for longer, reducing loss through decay, and allowing foodstuffs to be distributed over greater distances.

In order to gain the consumer's confidence, producers have endeavoured to preserve the original appearance of foods which are inevitably changed during preparation. Usually they achieved this with the help of colourings. Subsequently, it was the texture that they tried to preserve. They began to make fundamental changes to the look of the original foods, with the intention of improving their texture and taste. Emulsifiers in particular would give the foods a fondant texture which was not originally present. Although it does not itself constitute a flavour enhancer, the fondant does alter the taste of the product. At the beginning of the 1960s, this complete remodelling of foods went beyond reasonable limits; since then there has been a continuing decline in the use of additives.

The delicacy of monitoring equipment and of methods of medical and nutritional analysis has actually shown that not only were many additives highly toxic in the first instance, but that through synergy they could have a secondary toxicity. This reappraisal has given rise to some extreme attitudes which deny industrial food production any value and attribute to it an exaggerated harmfulness, as well as controversial claims of damaging effects (hyperactivity and lack of concentration in school children, for example).

Without additives, our everyday food would be considerably more expensive and less healthy; however, their scope, and the quantities in any product, must be strictly limited. Additives are a typical example of an invention which has been used too freely initially and has then had to be restrained.

On industrial food production, Dr Brouardel's famous diatribe in the late 19th century stated the dangers clearly: 'When a man has had his breakfast using milk preserved with formic aldehyde, when for lunch he has eaten a slice of ham preserved with borax and spinach made green with copper sulphate, and when he has washed it down with a bottle of wine treated with fuchsia or too much lime, and he has done this for twenty years, how on earth can you expect this man to still have a stomach?' Despite some undeniable excessiveness in the use of food additives, the fact nonetheless remains that in the last century, great progress has been made.

An example of the nutritionists' state of uncertainty about food additives, which has lasted about thirty years, is given in the studies on **nitrites**, potassium salts used for preserving food. From 1958, nitrites were thought to be capable of causing cancer of the digestive system in consumers who regularly ate preserved foods. It was only in 1988 that nitrites, valuable in the preserving of food and protection from botulism, were declared safe.

Food irradiation

Röntgen, Becquerel, 1896

It was as far back as 1896 that the famous German physicist, Wilhelm Konrad von Röntgen, and the no less well-known Frenchman, Henri Becquerel, formulated the theory that it would be possible to preserve foods using radiation, for X-rays could destroy the ordinary sources of decay: **bacteria, moulds** and **insects**. This idea was greeted with reserve, given that the food industry had only began to think about long-life preservation, and that no one knew the ultimate effects of such a radical method of stopping food decay. The 1970s saw a revival in interest in Röntgen and Becquerel's theory, and at the end of the 1980s large plants were set up for the irradiation of food.

It should be noted that irradiation does not seem to suit all types of food. Indeed, it caused fats to turn somewhat rancid, and so the taste is altered as a result of biological change. It is possible that irradiation may benefit the proliferation and mutation of certain bacterial cultures.

Food irradiation is still controversial. It is illegal in the UK and USA but legal in the Netherlands: other countries, such as Japan and Canada, have allowed distribution of irradiated foods.

Freeze-drying
Bordas and d'Arsonval, 1906

Freeze-drying or lyophilization is the process of using cold to dehydrate a food. The ice is removed by sublimation during rapid freezing. It is not new; in fact it dates back a thousand years, for it is known that the Incas of Peru used it to preserve their foods before Spanish colonization. It would be more appropriate, therefore, to define it as a re-invention. In 1906 the Frenchmen Arsène d'Arsonval and George Bordas became interested in a machine to dry meat within a vacuum. The machine, dated 1893, was not their invention, but that of the German, Karl P. G. von Linde. It was presented at the Universal Exhibition in 1900 and was bought by d'Arsonval; it could freeze things extremely quickly, at −140°C, almost instantaneously. D'Arsonval used it from 1900 to puzzle his friends when

he presented them with pieces of 'vitrified' steak.

Freeze-drying only began to be widely used forty years later, mainly for **preserving biological medical products**, vaccines, tissue, bacterial cultures, blood plasma and pharmaceutical products.

Being a relatively expensive process due to the equipment required, until the end of the 1980s it was of interest only to a few sectors of the food industry, for example those dealing with spices, coffee, powdered meat and vegetables (particularly for soups), and fruit juices. Much research was necessary in order to obtain foodstuffs of satisfactory flavour, which was facilitated by the introduction of microwave heating, vibrators and variations in pressure during the process.

Instant coffee
Nestlé, 1938

Following the suggestion of the Brazilian Institute of Coffee, in 1930 the Swiss firm Nestlé began to research the possibilities of producing a **dried coffee powder** which, when mixed with boiling water, would produce coffee instantly. This complex research lasted eight years. The coffee powder had to be soluble and could not just

form a deposit at the bottom of the cup; the taste had to be good, and the container had to be not only airtight, but filled under vacuum. A satisfactory product was not arrived at until 1938.

Travellers, workers and the idle gave it a welcome which has never waned: this was **Nescafé**.

Key-opening tin can
Ousterhoudt, 1866

Until 1866, tin cans required a can-opener, a gadget which could sometimes be difficult and risky to use, and which the workers did not always have with them at mealtimes on the worksite. That year the American, J. Ousterhoudt, thought of providing cans with a key which would enable one surface

to be rolled up along a groove cut at the canning factory, sufficiently shallow so as not to weaken the package but deep enough to allow the chosen side of the can to be detached. This process quickly became a great success.

Low-cholesterol egg

May, 1988

It was by changing his **battery hens'** feed that Paul May, manager of Rosemary Farm, South Maria, in California, managed to produce eggs which show an average cholesterol level of 125 mg as opposed to the usual 280 mg (figures confirmed by the Californian State Department of Food and Agriculture).

Margarine

Mergé-Mouriès, 1868

Originally what was called margarine was a mixture of purified melted tallow, butter, milk and vegetable oils. It was invented by the Frenchman, Hippolyte Mergé-Mouriès who was inspired by a competition launched by Napoleon III to find a product to replace butter.

According to modern classification, it was really a **blended margarine**, since it contained animal fats; the same classification is used today to distinguish purely vegetable margarines. Margarine is basically an oil-in-water emulsion which is given the consistency of butter by intense mechanical action — churning, then blending and lamination.

> The calorific value of margarine is the same as that of butter — 780 calories per 100 g; it contains no natural vitamin A, but both this and vitamin D are added in the UK and the USA.

Mechanical cow

Agrotechnic, 1987

A 'mechanical cow' is an apparatus which produces a liquid similar to **milk** from **soya**. Entitled Agrolactor and constructed by the French firm Agrotechnic, it measures about 4m³. The soya beans are poured into a hopper, crushed, pasteurized by being brought to 100°C, then rapidly cooled to 4°C. The product, which is rich in proteins and fats just like cows' milk, has the advantage of being easily digestible by both those individuals and ethnic groups, especially Africans, who do not have the enzyme lactase which is necessary for the assimilation of cows' milk. Moreover, it is richer in polyunsaturates so does not increase food cholesterol build-up. One kilo of soya and water is enough to produce 8 l of milk.

Milk chocolate

Peter, 1875

Though known in Europe since the end of the 17th century, until the end of the 19th century, chocolate was consumed mainly in drink form with sugar added. But it was customary to make it smoother by adding milk when it was prepared as a drink. It was this that gave the Swiss, Daniel Peter, son-in-law of the chocolate-maker F. L. Cailler, the idea of making solid milk chocolate. Production began in 1875.

Pasteurization

Pasteur, 1863

Pasteurization represents one of the biggest revolutions in the food industry. Also known as **'directed and controlled fermentation'**, it consists of destroying pathogenic and other bacteria in a food by heating it to a temperature usually less than 100°C, followed by rapid cooling. Asked by wine growers to determine the causes of fermentation in alcoholic drinks, Louis Pasteur established that it was due to specific **yeasts** which could be destroyed by heating to 55°C. Pasteurization only allows limited conservation but it does not alter the taste of foods or drinks. Its main advantage is in considerably extending the storage time of milk, cider, beer, fruit juices (and in some cases, wine as well). It allowed the industrial distribution network to be rapidly expanded and retailers' stock-taking to be regulated.

Plastic bottle

Kahlbaum, 1888

The first bottle capable of sustaining knocks without breaking was invented in 1888 by the Swiss chemist, G. W. A. Kahlbaum, who made it out of a type of polymer called **methacrylate**.

Refrigeration

Carré, 1857; Linde, 1873

It has long been known that food can be preserved in the cold. For centuries snow was used to keep certain animal foodstuffs fresh, as proven by the snow tanks discovered in the cellars of Hadrian's villa in Rome. Around 1660 the Italian, Zimara, recommended a **mixture of snow and saltpetre** as a refrigerating substance. Then it was discovered empirically that the **rapid evaporation** of heated **brine** absorbed heat. This latter process was derived from that of Zimara and from the Turkish porous clay water pots which chill the water they contain by the evaporation of the water that seeps out. In the 18th century it culminated in the first attempts at **controlled refrigeration**. The first step towards modern refrigeration was useful to confectioners, who could then sell frozen sorbets.

At the beginning of the 19th century, the progressive — and anonymous — development of this technique enabled the production of **artificial ice** in blocks for the first time. It was made in closed metal drawers which were filled with purified water and plunged into baths of brine; pipes carrying steam through the brine encouraged rapid evaporation and at the same time the water froze in the drawers. In the 1830s the mastery of steam machines and, later, of electricity, enabled the ice to be industrially made. This was commercialized to the benefit of the public, and huge underground cold-rooms began to be built in towns.

In 1857 the Frenchman, Ferdinand Carré, invented **refrigeration by compression** which marked the beginning of modern refrigeration. The principle was to distribute a volatile liquid, in this case **ammonia**, by using pipes around the area to be refrigerated. This technique rests on the fact that a body which vaporizes absorbs heat. The vaporization can be accelerated by producing a vacuum above the body (Carré based the industrial method of making ice on this). Carré's vacuum was obtained by using a compressor. Of course, at the beginning his apparatus was immobile. In 1873 the problem of movable refrigerators was solved by the German, Karl von Linde, who first used **methyl ether**, but this was dangerously explosive; consequently Linde returned to the ammonia advocated by Carré. From that time onwards railway wagons and ships could be equipped for the massive transportation of perishable goods.

In the United States, tons of butter were conveyed over large distances in wooden wagons refrigerated by ice placed on troughs filled with sawdust. After 1873 it became possible to transport huge tonnages of frozen meat across the Atlantic. The first **refrigerated ship** was the *Paraguay* which, from 1877, carried meat from Argentina to France; it was Carré who organized the equipping of the ship.

It is possible, but not established, that the principle of refrigeration was invented in China. Indeed, in the 14th century, Marco Polo brought back the secret of making milk sorbets from there, which may have been based on the principle of the evaporation of brine. The Chinese, who used brine a great deal in food preservation, could not have failed to notice its refrigerating properties.

Saccharin

Remsen and Fahlberg, 1879

The discovery of the first of all the **synthetic sweeteners**, 2-sulphobenzimide or saccharin, is due to the Americans Constantin Fahlberg and Ira Remsen, who made it during their research on the synthesis of a derivation of tar. It was associated with an invention, which was the use of the substance as a **food sweetening agent**. The American medical authorities, however, only marketed it under a pharmaceutical label and on prescription until 1938. Thereafter, the Food & Drugs Administration authorized it to be added to industrial foods on the condition that this was controlled. Its use grew rapidly in the industrial production of drinks and confectionery, given that it had no calorific value, which suited those on weight loss diets, nor carbohydrates, which suited diabetics. The banning of **cyclamates** increased its consumption further but its use in processed foods is limited by the fact that it is destroyed by heat.

Silo

Hatch, 1873

Silos, or **cylindrical airtight foodstores** seem to have existed since the first century in the Middle East and in China, and the Romans built them with remarkable technological expertise. They were used either as **grain silos** or **fodder silos**. Thus it cannot be said that their reappearance at the end of the 19th century constitutes an invention as such. However, the vertical silo had disappeared from the Western world when in 1873 the American, Fred Hatch, tried to build one in McHenry County, Illinois. It is not clear whether his attempt was successful; nevertheless, it marked the reappearance of the silo in agriculture.

Chemistry and Physics

From the synthesis of acetylene to the production of metallic xenon, chemistry and physics have produced a number of ideas since 1850. Dirac's invention of antimatter constitutes one of the most prodigious intellectual tools to aid the understanding of matter since the beginning of the history of physics. Furthermore, J-M. Lehn's cryptans stand as one of the most masterful inventions in chemistry for they have already begun to revolutionize the cosmetics industry and pharmacology as well as the industrial dyes industry and extraction chemistry. It is impossible yet to evaluate the impact of monocrystalline superconductors on the whole of the electronics industry.

It is likely that the biggest changes to take place in the next few decades will essentially be inventions in chemistry and physics. Commercial interest too has undeniably contributed a great deal to the wealth of invention in these two areas. It must be stressed that, contrary to common opinion, the greatest inventions are the fruit of fundamental research which is carried out far from any commercial concerns. They can be compared to brilliant flashes of understanding which appear in solitude and anonymity, which are then adopted by industry and carried to the point of practical completion.

Acetylene (synthesis of)

Morren, 1859; Berthelot, 1862

In 1959 Marcel Morren, the dean of the science faculty of Marseille, published an article in the 'Reports of the Academy of Science' in which he reported that having activated an electric spark in a glass ball which contained some **carbon** electrodes and **hydrogen**, he had obtained a '**carbonized hydrogen**', the nature of which he had not yet established. In fact it was **acetylene**. Three years later, Marcelin Berthelot, already famous for his invention of the **synthesis of alcohol** (see *Great Inventions Through History*), repeated the same feat. However he baptized Morren's glass ball 'electric egg' and he knew that he had synthesized acetylene. There was one difference between the two experiments; Morren had made a **discovery** and Berthelot, an **invention**. Jean-Baptiste Dumas, who was president of the Academy and a chemist of great renown, drew Berthelot's attention to what Morren had done before him. Berthelot replied that Morren had not been able to verify his production of a carbonized hydrogen. In fact, Morren had verified it by a **spectral analysis** of the gas obtained, but he was not up to contesting the matter with Berthelot; he stepped down and history has forgotten his name.

> The injustice committed by Berthelot and posterity regarding Morren is one of the worst in the history of inventions. The historian, Jean Jacques, tried to make amends in 1987 in his work, **Berthelot**, *Autopsie d'un Mythe*.

Acrylic paint

Reeves Ltd, 1964

Acrylic paint is made of a solution of **acrylic vinyl copolymers** and is much easier to handle than oil paint. It was invented in 1964 by the British firm Reeves Ltd and has been successful in the art world where it has resulted in a new technique.

Adhesives

Anon., c.1850; Certas, c.1958; Eastman, 1961; European Society of Bonding, 1961

It would require considerable space to give an account of all the inventions involving adhesives, given that these products have been and still are the subject of international research which is incessantly punctuated with improvements. Nevertheless, the start of the age of adhesives is considered to be around 1850 when it was suddenly realized that rubber was accompanied by a later invention: glue based on a rubber solution; once the **solvent** evaporates, the rubber forms a more or less sticky adhesive layer. This principle remains to this day; a volatile solvent is added to a substance with an adhesive base. Thus the range of organic, animal and vegetable glues grew.

The progress made in **organic chemistry** in the first half of the 20th century culminated in the invention of glues based on polymers (molecules built up of many smaller units, or monomers), the first of which was made in France in 1958 by the Certas Company. New progress was made in 1961 when after ten years of research the firm Eastman-Kodak developed a **cyanoacrylate monomer** called Eastman 910, which had remarkable adhesive properties. In the same year the European Society of Bonding produced Araldite, also with a polymer base.

Formulae for adhesives will multiply in the future in relation to their specific uses. Despite some claims to the contrary, no universal glue actually exists; every glue is designed not only for particular materials but also for specific conditions of use. Broadly speaking, the adhesivity is determined by the polymerization (combination of several molecules to make a more complex one) of the adhesive, which is pro-

duced by the effect of either humidity, the absence of oxygen or of heat. This distinction is not only based on the physical and chemical characteristics of the adhesive, but also on the fact that an adhesive sticks well to a surface according to the nature of the surface. In fact, the aim of the adhesive is to create a molecular bond which is as near as possible to the actual structure of the material to be glued, but which is, in theory, impossible. Indeed, the finest polished surface cannot end up perfectly joined to the adhesive–substance interface. The adhesivity therefore is really the adhesive's relative **capacity to penetrate** into the irregularities of the material (wood, for example), or else it is due to **molecular attraction**, a fundamental phenomenon which did not begin to be understood until around 1960 and which allowed metal glues comparable to solders to be made.

Aerogels

Kistler, c.1932

Gels, which have loosely linked structures and consist of colloidal substances swollen by a solvent, are characterized by the fact that the **polymeric chains'** attachment of each other is temperature-dependent. They have always interested physicists and chemists because above a given temperature they become liquids and below it, they become like solids. Another feature present in gels is that they dehydrate when exposed to air. Around 1932 the American S. S. Kistler, from Stanford University, had the original idea of replacing the fluid solvent with a **gas**; he then extracted the gas under pressure at a high temperature; the structures of the polymeric chains remained

unchanged, but the gel became 98% porous, with the solvent having been replaced by air. This is what Kistler called an aerogel. During the 1960s a French method of rapid manufacture of aerogels allowed their everyday use in the **particle detectors** in high-energy experiments. The study of them has shown that they are remarkable **thermal insulators** (i.e. do not conduct heat), a hundred times more effective than the densest silica glass. At the end of the 1980s research was being carried out in numerous laboratories around the world on the physical, optical and acoustic properties of aerogels.

Antimatter

Dirac, 1928

The fundamental concept of antimatter was proposed by the mathematician P. A. M. Dirac four years before the American Carl Anderson discovered the first particle of antimatter, the **positive electron or positron**, in 1932. This discovery was derived from a study of relativistic equations which led Dirac to postulate the existence of states of matter which are energetically negative. This profound idea seemed almost unacceptable because although it is possible to imagine a negative charge, such as for the electron, it is difficult to conceive of an 'anti-energy'. It was however a particle symmetrical to the electron that Anderson observed on a photographic plate which showed the paths of particles of cosmic origin after passing into a gas counter built at the Californian Institute of Technology (Caltech). Anderson had not heard of Dirac's theory, of which only a few high-level mathematicians were aware.

Since then, the concept of antimatter has continued to be the object of research in physics and astrophysics. Dirac had also postulated that a **positron** and an **electron** would tend to unite according to Coulomb's law (which states that opposites attract) thus forming a combination called **positronium** which would, however, have a very short life, disintegrating to produce **gamma rays** (very high energy radiation), after between 10^{-7} and 10^{-10} seconds. This too was verified experimentally. Other verifications followed. In 1955 a team of physicists under the leadership of Owen Chamberlain and Emilio Segrè observed that, according to mathematical predictions, when very high energy **protons** collided they would produce anti-protons and would release an enormous amount of energy. At present, symmetrical (anti-)particles for nearly all known particles have been discovered.

The next stage consisted of postulating the existence of antimatter made of various antiparticles in space. This hypothesis even included the possibility that **quasars**, **pulsars** and **neutron stars** might be antimatter centres isolated in the universe. At the end of the 1980s this hypothesis had not yet been verified, and all theories of cosmology found it impossible to explain the existence of separate masses of matter and antimatter in the universe. Indeed, if within the context of the original explosion or Big Bang, one agrees that matter has one sole origin, it is impossible to see how matter and antimatter came to be separated. Besides, the **gravitational repulsion** between matter and antimatter is incompatible with the General Theory of Relativity. Some experiments on **mesons** and **anti-mesons** have nevertheless shown that the gravitational relations between matter and antimatter would be the same as those which govern the relations between matter and matter.

Artificial diamond

Hannay, 1878(?); Moissan, 1904(?); Allmäna Svenska Elektriska Aktiebolaget, 1930–53; Norton International Inc. and General Electric Co., 1949–55; De Beers, 1958

The history of the invention of the artificial diamond is confusing. The first person to have the idea and to try making it was the Scotsman James Ballantyne Hannay who in 1878 began some experiments aiming to produce artificial diamonds based on animal oil, paraffin oil and lithium in a steel cylinder. Under high pressure the fats vaporized and reacted with the lithium to form carbon, which remained fixed to the inner wall of the cylinder. Hannay made 80 attempts in two years; only three tubes resisted the pressure. In 1880 Hannay reported that he had found tiny crystals of pure carbon of which the density matched that of diamond. He sent these crystals to the official mineralogist at the British Museum in London who declared that they were real diamonds.

In 1896, Henry Moissan, the inventor of the electric oven, also tackled the problem and he resolved to obtain artificial diamonds from charcoal, iron and graphite, the last being used as a coating for the container. He put the powder mixture into his electric oven and turned the temperature up to 4000°C. The steel melted completely and with it a large part of the charcoal; then he plunged these fused materials into cold water. One characteristic of molten iron is that it expands on solidifying; Moissan's reasoning was that the sudden cooling would harden only the outer casing of the fused nodule, while a magma would remain inside which would also try to solidify and so to expand, but which would meet with considerable resistance in the form of the casing. An extremely high pressure would therefore develop in the nodule and Moissan hoped that it would result in the formation of diamonds. He found out that, in fact, once the ingots had cooled down and had dissolved in acid, some crystals had formed inside which were mixed with graphite.

These two experiments seemed valid; moreover, that of Moissan was confirmed by Sir William Crookes who ascertained that Moissan's method had resulted in the formation of diamonds. He reported also that the industrialist, Sir Andrew Noble had himself obtained diamonds by exploding cordite inside steel cylinders, creating a pressure of 15 000 atmospheres at a temperature of 4000°C. Other experiments subsequently verified the validity of Moissan's method.

However, in 1928, after 20 years of research, Sir Charles Parsons who invented the steam turbine announced that he did not believe that Hannay and Moissan had made real diamonds, but instead a very hard crystal capable of scratching glass. If Hannay's diamonds, which were still available for analysis at the British Museum, turned out to be real diamonds when studied using X-ray diffraction (which was

not known in the time of Hannay and Moissan), in Parsons' opinion, that still did not signify that Hannay had made them! Thus Parsons threw doubt onto the work of Hannay and Moissan but did not shed any light on the actual question.

In 1941 the American P. W. Bridgeman, pioneer of studies of **high pressures**, set up a collaboration of research with three American firms to study the manufacture of artificial diamonds under very high pressures. The war interrupted his work which was taken up again in 1949 by the Norton International Inc. and the General Electric Co. independently. In 1955 the General Electric claimed to have succeeded in their research but their disillusionment was great when the Swedish firm Allmäna Svenska Elektriska Aktiebolaget announced that they had pursued their own work on the project since 1930 and had achieved their goal in February 1953.

Once again, the Swedish claim gave rise to scepticism, all the more because the team which was claiming to have succeeded in making diamonds, led by the engineer B. von Platen, maintained that they were interested in jewellery stones, not in industrial diamonds. Now, in order to make quality jewellery diamonds, a pressure of 60 000 atmospheres is needed, which the technology of the age did not seem to allow. And yet the Swedes proved their assertions to be valid. Bringing the latest mechanics and metallurgy into play, they had the idea of enclosing blocks of **graphite** and a **catalyst** such as **nickel, cobalt** or **iron** in a matrix of **tungsten carbide**. Enormous pressures were applied to this which were generated by a set of convergent truncated

Perhaps Parsons' scepticism was inspired by the fraudulence of Henri Lemoine, an adventurer who persuaded the president of De Beers, Sir Julius Wernher, that he had succeeded in making industrial diamonds. Lemoine gave himself away by giving some specimens to Wernher which were in fact diamonds from actual De Beers' mines. The extreme resemblance of his 'products' to the natural diamonds meant that he was discovered and had to spend six years in prison.

pyramids activated by a press. This system has since been developed by General Electric and is known under the name Belt System. In 1958 De Beers of South Africa, the mining and rough diamond merchants, announced that they too had arrived at convincing results.

The manufacture of synthetic jewellery diamonds is not impossible, but is hardly profitable. In 1970 General Electric succeeded in producing an artificial diamond of the highest quality and the appreciable size of 1 carat. This was exhibited at the Smithsonian Institution in Washington, but its cost price was considerably higher than that of a natural stone. All of the 140 million carats which are now made by South Africa, Sweden, Ireland, the USSR, Japan and the United States consist of diamonds for industrial use only (millstones, record player cartridges, boring heads etc.).

Aspirin

Gerhardt, 1853; Kolbe, 1859; Riess and Stricker, 1876; Hoffman, 1899

Widely known among chemists and pharmacologists (see *Great Scientific Discoveries Through History*), aspirin (acetyl salicylate) still remained to be prepared in a digestible and convenient form. The modern age of the aspirin began in 1838 when the Italian, R. Piria for the first time isolated pure salicylic acid, an extract of methyl salicylate, which he himself had taken from willow tree bark. But the modern inventor of aspirin is really the Frenchman, Charles Frédéric Gerhardt who in 1853 obtained the first acetyl salicylic acid by treating sodium salicylate (from silver birch tree bark) with acetyl chloride. Numerous knowledgeable or historical publications omit Gerhardt's name for it was overshadowed by that of the German, Hermann Kolbe. The reason for this is that neither Piria's salicylic acid nor Gerhardt's acetyl salicylic acid were obtained in a **stable form**, and their industrial production was difficult due to the short supply of willow and silver birch.

Kolbe's genius took root in the great tradition of **industrial synthesis** which characterized German chemistry in the 19th century. In fact, Kolbe obtained salicylic acid in 1859, without recourse in any way to a natural product. He heated a mixture of **phenol** and a hot aqueous solution of **sodium hydroxide** in an autoclave; at 130°C it forms sodium phenate which is dehydrated by heat and then cooled and, since 1874, subjected to an injection of carbon dioxide gas under high pressure. This product is heated again for several hours at 150°C, then cooled: **sodium salicylate** is obtained. After this has been dissolved in water, then acidified, pure salicylic acid precipitates and is collected by centrifugation, dried and finally purified by recrystallization and sublimation. The Kolbe process allows salicylic acid to be manufactured industrially.

Salicylic acid was shown to be poorly tolerated by the stomach, so in 1876 the Germans, Peter Theophil Riess and Salomon Stricker, took up the Gerhardt process again and altered it; they advocated treating the salicylic acid with acetic anhydride. It was the German, Felix Hoffman, who was the first in 1899 to make aspirin (**acetyl salicylate**) in the stable form in which it has thenceforth been known. This he did in the laboratories of the Adolf von Bäyer firm at Elberfeld.

The first uses of salicylic acid, whose medicinal properties are still being explored to this day, were of interest to the dyeing and textile industries for it produces a violet dye when heated to 200°C with ferric chloride. One of its derivatives, **salicylanide**, is also a **fungicide** used to protect cotton threads.

Borazon

Wentorf, General Electric, 1957

Boron nitride or BN is a very hard material which was obtained at the end of the 19th century by the Frenchman Henri Sainte-Claire Deville and the German Friedrich Wöhler, by bringing boron to a very high temperature in a stream of ammonia. It consists of a white, bulky powder which, when it reaches its very high (3000°C) fusion temperature, makes an excellent lubricant for high-pressure bearings. It has a **hexagonal structure** similar to that of graphite. In 1957 the American R. H. Wentorf of the General Electric Company obtained a different form of boron nitride which this time had a **cubic** structure, harder than diamond, which was a result of the use of very high pressure. Its commercial name is borazon.

Butyl rubber

Thomas and Sparks, Exxon, 1937

The discovery of **rubber** in the 18th century, then the discovery of its **vulcanization** (treatment with sulphur) by the American Charles Goodyear in 1839, had created a huge industrial field which for a long time remained dependent on the plantations of the latex-producing hevea tree (*Hevea brasiliensis*). As usual, the industrialists strove to free themselves from this dependence which involved firstly an economic dependence and then considerable transport costs, the latex having to be imported from tropical countries such as Brazil primarily, then the Philippines and Malaysia, as well as Ceylon, where plantations of heveas had been planted; so chemists started researching synthetic rubber. Since the 1930s it was known that natural rubber was composed of 94% **polyisoprene**, the remainder being shared among the fatty acids, proteins, sterols and natural stabilizers; it was therefore possible to synthesize it from oil, which began to be done in various ways in Germany, Great Britain, France and the United States. In the United States, the firm Exxon obtained a petroleum by-product, **butylene**, from which it was possible to extract one of the principal components of rubber, **butadiene**. But the **synthetic elastomer** (material with elastic properties) thus obtained did not possess the characteristic strength of natural vulcanized rubber: a means of vulcanizing it remained to be found.

In July 1937 the Americans Robert Thomas and William Sparks from Exxon tested a molecule of their own invention, a **polymer** of **isobutylene** and **boric fluoride** base, which seemed to them to be likely to have the same effects as those of sulphur and heat on natural rubber. According to them, it should be possible, by adding a certain percentage of butadiene to their polymer, to achieve the supplementary bonding which would capture the sulphur and thus assure the vulcanization. The completion of the industrial process took a long time, for it was only at the beginning of 1941 that Exxon installed the first pilot factory for the production of butyl rubber.

Cryptans or electrides

Lehn, 1976

The name cryptans is derived from 'crypt' or cave, and is given to synthetic crystalline materials which can capture electrons, which act in a complex manner inside these traps, lending unusual chemical, optical or electronic properties to the cryptans.

Cryptans were invented in 1976 by the Frenchman Jean-Marie Lehn, for which he was honoured with the Nobel Prize for physics in 1987, together with the Americans Donald J. Cram and Charles J. Peterson, who developed Lehn's invention and gave it a third name, **crown ether**. This is self-explanatory due to the fact that the cryptates consist of two circles of atoms superposed, each circle itself consisting of ether molecules which are composed of a carbon structure to which atoms of oxygen and hydrogen are attached. It is the interstices between these structures which act as traps. The capacity for attracting the electrons is due to the fact that at the centre of each cryptate there is an atom of one substance, **caesium** for example, which has lost an electron and which therefore has a positive charge. The principal difficulty in the fabrication of cryptans lies in the risk that the electrons, interacting with the atoms of the traps, might cause the destruction of the whole thing; this difficulty can be prevented by the perfect purification of the components by the manufacture and storage of the cryptans at very low temperatures.

It should be added that at these temperatures, approaching −263°C, the cryptans frequently display a curious characteristic in becoming **antiferromagnetic**, that is, half of their electrons align themselves spontaneously with their spins in one direction, and the other half in the opposite direction. This characteristic is absent however in cryptans which have a **rubidium** or **potassium** base, though it is not known why.

Cryptans constitute one of the major inventions in the field of chemistry in the 20th century. Their uses are very varied: in medicine for example, cryptans can be used as **tracers** in certain diagnoses, and in the case of radioactive contamination they can capture radioactive **strontium** or **cadmium**. In industry, Rhône-Poulenc already use them for making **ceramic polymers**, and there is a promising application envisaged in the field of energy, for some cryptans can in fact collect **solar energy**, behaving somewhat like chlorophyll (the green pigment in plants which absorbs sunlight). The solar energy would be captured in the following way: the cryptans' electrons are loosely connected; the impact of a **photon** ('packet' of light energy) is therefore enough to dislodge them and subsequently a weak electric current is generated.

Diamond film

Rabelais and Kasi, 1988

It was by introducing low-energy carbon ions (C^+) to a very advanced vacuum $(10^{-10}$ torr) that the American, J. Wayne Rabelais and the Indian, Srinandan Kasi managed in 1988 to make layers of very pure carbon, like diamond. These had the structure of **carbide**, and adhered tightly to the surface which supported them. Being very poor conductors, and therefore very good insulators, as well as good heat conductors, with a toughness similar to that of diamond these films are used for a large variety of technological purposes, not only as insulators but also as **doped semiconductors** and in **industrial grinding**.

Explosives

Sobrero, 1846; Schultz, 1860; Nobel, 1862, 1867, 1876 and 1887; Wilbrand, 1863; Vieille, 1884; Linde, 1895; Du Pont, 1934; Akre, Cook, 1955

The first modern explosive, **nitroglycerine**, is a result of research (albeit unsuccessful) on textiles; it was invented in 1846 by the Italian, Ascanio Sobrero. Dynamite, which is absorbed by sticks of porous silica or kieselguhr, or else by sticks of charcoal, was made for civil engineering by the Swede, Alfred Nobel; it is really merely an adaptation of nitroglycerine. It was still for the purposes of civil engineering that in 1867 Nobel was to perfect the **primary dynamites**, which were less dangerous to handle and less prone to freezing, but which reduced their detonating power; instead of being based only on pure nitroglycerine, they included **polyhydric alcohols** and sugars mixed with **wood pulp**.

While working in the area of dyes, the Swede J. Wilbrand made an important step forward in 1863 in **thermochemistry** by inventing TNT, or trinitrotoluene. TNT was the result of the nitration of toluene, a distillate of crude oil. In the meantime the Prussian, E. Schultz, pursued some fruitful research which he had begun in 1860; his aim was to complete an agent of **propulsion** for artillery missiles. His process was complex; it consisted in effecting the nitration of pieces of wood, then washing them to take out the excess nitrate and impregnating them with barium and potassium nitrates, which supplied oxygen to the fragments of wood containing traces of nitrates. Schultz was not satisfied for his propulsion was too rapid for artillery shells and for most guns; it was hardly even suitable for advanced shotguns. But this explosive, which was the first to be made for specifically military purposes, served as a starting point for research into **propellants** in the field of astronautics and was to lead to **propergols** at the beginning of the 20th century.

In 1876 Nobel, who was specializing in the manufacture of explosives, patented **explosive gelatines** and **gelatinous dynamites**, which would later be developed into **plastic** explosives; these products were firstly by-products of the manufacture of nitroglycerine, and, secondly, more fluid gelatine explosives which were combined with a solid substratum and so less dangerous to handle. From the mass of nitrocellulose that is dissolved in a nitroglycerine bath, 7–8% of a gelatinous product is actually obtained: this was the starting point for these new explosives, the **high explosive** sort, for they exploded at a speed of 8000 m/s, compared with the 4000 m/s of nitroglycerine.

New progress was made in 1884 by the French chemist Paul Vieille, who invented **smokeless gunpowder**. The process consisted of dissolving nitrocellulose in a mixture of ether and alcohol until a gelatinous mass was obtained, the cellulose particles of which burned without making any smoke. This gelatine was then dried and a product similar to corn was obtained, which could be cut up into small pieces to be used.

In 1887 Nobel also invented a revolutionary product from nitrocellulose and nitroglycerine with a nitrogen base, which was called **ballistite**. This was precisely the propellant that Schultz had researched a quarter century previously. The English however refused to recognize his patent and made the same product for their own use under the name **cordite**.

The improvement of explosive power allowed work on a previously unknown scale to be undertaken, such as the Mont Blanc Tunnel which was dug out under the Alps between France and Italy from 1857 to 1871. Such exploits were facilitated further by the invention of the **Bickford fuse** by the Englishman William Bickford, then by the completion of **remote electric ignition**, by the American H. Julius Smith. The transport and handling of explosives caused problems; so the oxidizing agent and the fuel began to be taken directly to the site where they were mixed and then ignited from a distance. Potassium chlorate and nitrobenzene were combined in this way, shortly before the explosion. In 1895 the German Karl von Linde invented a process which consisted of throwing a porous fuel, such as charcoal, into liquid oxygen which was made to explode from a distance; this was a new kind of explosive, among several others, called **liquid oxygen combustible**.

Military use of the new explosives did not really begin until the Russo-Japanese War of 1905, and since then it has not stopped. In 1907 the firm Nobel made a very important improvement by finding a way of preventing dynamite from freezing. Until then at temperatures of around 0°C, dynamite had been virtually useless, and careless attempts to warm it had resulted in numerous serious accidents. The invention consisted of blending a fifth to a quarter of **trinitrotoluene isomers** or TNT into the nitroglycerine. An isomer consists of molecules of identical form but different structure and therefore different properties. Shortly afterwards, this was replaced by another invention which consisted of incorporating a solution of nitrated **sugar**, which was less sensitive to the cold, into the nitroglycerine. In 1911 there was another improvement with the discovery of a way to make a polymer of glycerine, diglycerine. This, when nitrated, produced **tetranitroglycerine**, which, when mixed with nitroglycerine considerably reduced the freezing point. Finally, in 1925 there was another step forward in the development of dynamite which could be used at low temperatures, due this time to the mass production of **ethylene glycol**. This allowed the use of nitroglycerine to be discontinued, since the produce of the nitration of ethylene glycol, ethylene glycol dinitrate, had the same properties.

In 1934 the American firm Du Pont invented an entirely new explosive, Nitramon, which was composed of 92% ammonium nitrate, 4% dinitrotoluene and 4% paraffin wax. Nitramon differed in its great **shock resistance**, even to the extent of a bullet fired from a small-bore gun; thus

an element of previously non-existent safety entered into the domain of explosives. It exploded only with the help of a primer produced by the manufacturer. A more powerful variation was Nitramex which contained TNT and metallic ingredients such as aluminium. Its most spectacular use was for the destruction in 1958 of a rock which was making navigation hazardous between Vancouver Island in British Columbia and the Canadian coast. ·

Other considerable progress resulted from the invention of explosives based on mixtures of ammonium and fuel oil, as well as of gels based on ammonium nitrate and water. The former formula was found in 1955, inspired by the **carbon gunpowder-based explosives** invented in 1947 by the American Robert Akre, and was patented under the name of Akremite. The latter formula was made more or less simultaneously by the American Melvin Alonzo Cook, a former employee of Du Pont, and by several other chemists around 1958; it consisted in mixing together ammonium nitrate, TNT, water-based gelatine and colloids.

These later explosives, referred to under the generic name of plastics, are widely used because they contain very highly concentrated explosive products and their plasticity enables them to be inserted into cracks and handled comfortably since they can only explode with the action of a detonator. It is these explosives which, unfortunately, have led to the spread of terrorism.

The industrial manufacture of explosives was dangerous for a long time. In 1862, for example, the Nobel factory at Heleneborg in Sweden exploded, and the inventor's father lost his life. The first man to succeed in the controlled manufacture of nitroglycerine was the American George Nowbray who, in 1867, managed to make 500 tonnes of this product and transport it by rail and sea without one accident.

Ferrofluids

Rosensweig, 1968

Also known as **magnetic liquids**, ferrofluids represent one of the major inventions of the 20th century, after the fashion of liquid crystals, heterogeneous materials and composite materials. Invented in 1968 by the American Ronald Rosensweig, they are fluids which contain a suspension of infinitesimal grains of a magnetic material, such as **ferric oxide**, in the region of a millionth of a centimetre in size. Their uses

are extremely varied and at the end of the 1980s they were still not all known. A Japanese company, Matsushiata Electric Industry (Osaka) has produced a high-resolution printer capable of printing five pages a minute using a **ferrofluid ink**; small magnets make microscopic focused projections of ink onto the paper (8 dots to a millimetre). The American firm Ferrofluidics has made a **ferrofluidic lub-**

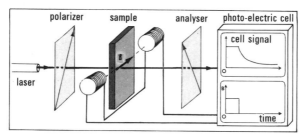

One of the most interesting facets of **ferrofluids** *lies in their optical properties. Thus, a ferrofluid placed in a magnetic field allows the passage of light, because its grains line themselves up in the field. But if the field is cut off, the ferrofluid slowly becomes opaque because the grains regain their former position. This property allows, for example, the viscosity of a fluid to be measured.*

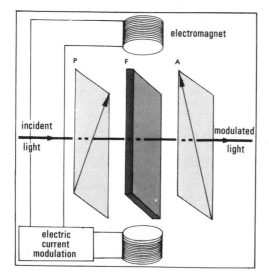

The optical properties of ferrofluids also enable the detection and measurement of a magnetic field, even if it is very weak. It is known that a ferrofluid exposed to a magnetic field allows the passage of light. If there is no field, it remains opaque. In the device shown, an example of fluid with a thickness in the region of 100 mm (1 mm = one millionth of a metre) is placed between a polariser P and an analyser A. If an electro-magnet controlled by a current modulator is added, like the one opposite, a light modulator is formed.

ricant using collars of magnetic liquid which hold themselves in place. It considerably reduces the wear on joints since it is waterproof as well. Another Japanese firm, Diesel Kiki, is equipped with an **inclinometer** which is ultra-sensitive to ferrofluids, measuring the heaviness at every point of the liquid. The US Air Force benefited considerably in 1987 with the manufacture of a **paint** which makes aeroplanes invisible to radar; it is based on ferrofluids and non-magnetic substances stopping the reflection of radar waves.

The refractive properties of ferrofluids (each grain, a micro-magnet, has a real magnetic moment which means that it refracts light) give rise to numerous optical applications, such as the measuring of the specific viscosity of a liquid placed between a polarizer and an analyser, illuminated by a **helium—neon laser**. Of course, ferrofluids can be used as very sensitive magnetic field detectors. A medical use of **injectable ferrofluids** can also be envisaged; given that cancerous cells have an affinity for ferric oxides, a compatible ferrofluid can be used for their detection. Given also that the magnetic particles absorb **X-rays** and produce a localized heat, the destruction of cancerous cells by **attenuated radiotherapy** should also be possible (research by Patrick Couvreur at the Laboratory of Pharmacy and Biopharmacy in Châtenay-Malabray). There are numerous disciplines within inorganic chemistry and the physics of magnetic colloids which are endeavouring to define the fundamental characteristics of ferrofluids.

Fluorescent paint
Switzer Brothers and Ward, 1933

It was in collaboration with the chemist Dick Ward that in 1933 the brothers, Joe and Bob Switzer, all Americans, invented fluorescent paint under the trade name Dayglo. It was a paint based on **zinc sulphide**, a substance whose atoms contain electrons which tend to change orbit under the influence of light energy; these electrons return to the original energy level as soon as the light disappears. This is the phenomenon of **electroluminescence**.

Free electron laser
Madey, c.1970

A conventional laser (see p. 32) consists of a device designed to produce a coherent beam of light. Its capacity depends on the wavelength of the light, which is dependent on the colour of the beam. Since the beginnings of the laser it was **modulation lasers** or lasers of variable wavelength which scientists were endeavouring to perfect, for such a laser could then be used for several purposes. To do this, the use of **dyes** was attempted. These would modify the colour and therefore the wavelength of the transmitting source; but the results were disappointing, given that this type of laser which nevertheless exists and which is even used in some laboratories, has only a low power and a restricted range of wavelengths. At the beginning of the 1970s the American physicist John Madey, of Stanford University, designed another type of laser based on the following principle: when a **charged particle**, such as an electron, is deflected by a magnetic field, it emits a photon; so by passing a beam of electrons across a circuit of magnets arranged in a certain order, the beam of emitted photons can be used to make a coherent beam. The wavelength of this beam depends therefore, either on the strength of the deflecting magnetic field, or on the energy of the electron beam, or on a combination of the two. This is what is called a free electron laser. There are considerable advantages to this type of laser because not only can it be modulated, but it transforms up to 20% of the energy produced into light, whereas conventional lasers only transform 1%.

At the end of the 1980s, the free electron laser was being perfected in the laboratories of several large American bodies, T.R.W., the Bell laboratories of American Telephone and Telegraph, Boeing, Vanderbilt University, Lawrence Livermore National Laboratory and Los Alamos National Laboratory.

The interest that these organizations show in Madey's invention can be explained by two factors. The first is that the free electron laser can be used by the programme nicknamed 'star wars', or Strategic Defense Initiative (SDI), because it shows satisfactory signs of atmospheric penetration, which is not the case with conventional lasers. The second is that this device is invaluable in **surgery**, because it enables operations of extreme precision and depth which have never before been attempted. It seems that all that is required is that the tissues are dyed, even thinly, to make them vulnerable to a laser of the same colour as that of the dye. Thus cancer cells can be tainted with a certain colour, usually with a dye for which they have an affinity, and then destroyed selectively by a laser which crosses over the healthy tissues without damaging them.

In 1988 there were only six free-electron lasers in use in the medical field, all in the United States. All worked in the range of 320 thousand millionths of a metre, just beyond **violet** in the visible spectrum, to 800 thousand millionths of a metre, just beyond **red**. These devices were expensive, since they cost about £1.2 million each.

Laser

Townes, 1951–54; Maiman, 1960; Javan, 1961

The laser is one of the best examples of an invention which follows from a theory. Its precursor is perhaps Albert Einstein, who from 1917 developed the theory of emission of **radiation** stimulated by **photons**. Einstein postulated that if there is a higher than normal number of electrons in a high energy level, L_1 (**'population inversion'**), an incoming photon of energy E, (where E is the energy difference between level L_1 and a lower level, L_2) will stimulate an electron to jump down to level L_2, emitting another photon of energy E in the process. Thus the incoming photon beam is **amplified** by **stimulated emission** of photons of the same energy (and hence the same frequency). The process only works efficiently if there are a lot of electrons in the high energy level to start with — the population inversion must be maintained by injection of energy.

The principle of the laser is based in the first place on the amplification of a beam of photons by stimulated emission of photons of the same frequency. Thus, a monochromatic beam of light, the 'amplification chamber' of the laser, will be intensified by the stimulated emission of the photons. In order to in some way contain the amplification, a **cavity resonator**, also called a Fabry–Pérot cavity, is used. This consists of two facing mirrors, perpendicular to the radiation, which reflect the photons back and forth; one of these mirrors has a semi-silvered surface, to allow a fraction of the beam of photons to escape. It must be noted that the amplification is only achieved if the frequency of the beam of

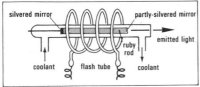
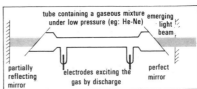

Diagrams of the ruby crystal and gas lasers. The two exactly parallel ends of the rubies are silvered, but one is slightly transparent to allow the emission of the beam of light. The helium-neon gas laser (on the right) has an emergent power of some milliwatts.

monochromatic light is exactly the same as that necessary to make electrons jump from a highly populated energy level to a lower energy level.

It was the American physicist, Charles Hard Townes, Nobel Prize winner for physics in 1964, who was the first, in 1951, to describe the possibility of short-wave amplification or **maser effect**, an acronym of Microwave Amplification by Stimulated Electro-magnetic Radiation. In 1954, Townes demonstrated this; if Einstein is the spiritual father of the laser, one could say that Townes is its 'practical' father.

The maser effect embraces all wavelengths; it is used in astronomy for the amplification of very weak radio waves. The laser effect is restricted to the optical domain, where Townes was also a precursor. In 1958, with his compatriot Arthur Leonard Schawlow, he described the optical maser, i.e. the laser, which is an acronym of Light Amplification by Stimulated Emission of Radiation.

The characteristic peculiar to the laser effect is that as the photons are emitted exactly in phase with the stimulating beam, they are **coherent**. This coherence means

A functioning laser: the beam emerging from the laser on the right is shown to be monochromatic by the prisms on the left.

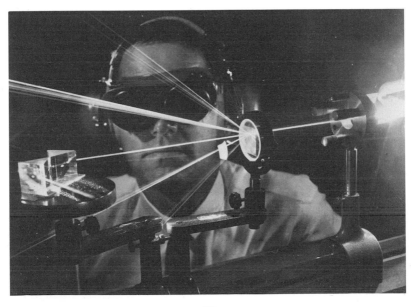

33

that the laser beam is **directed**, that is, non-divergent, in contrast to ordinary light which diverges and is said to be incoherent. Moreover, a focused beam can deliver considerable energy to a point of infinitesimal dimensions.

In 1960, the American T. H. Maiman created the first ruby laser. The following year his compatriot A. Javan made the first gas laser.

Since then, four large groups of lasers have been built (the free electron laser is omitted as it follows a different principle, see p. 31). First, the **sold material laser**, using rubies, neodymium glass and yttrium aluminium garnet or YAG; then, **gas lasers**, using ionized gas for ionic lasers, carbon dioxide, and metallic vapours; then, **chemical lasers**, which use the reactions of atomic fluoride on hydrogen or deuterium to produce excited molecules, hydrogen fluoride or deuterium fluoride, and also iodine lasers; finally, the **semiconductor lasers** where it is the doping (addition of impurities) of certain semiconductors which produces the excited levels. One group is different, the **dye lasers** which use a liquid centre, and which provide the transition to **free electron lasers**.

There are numerous uses of lasers, from metallurgy, for example, where they are used in cutting and soldering, to the measurement of surfaces and alignments, from telecommunications to photography, and from medicine to warfare. They have resulted in significant improvements in measuring instruments and instruments of observation, in the development of very precise **telemeters** (which function on the principle of radar which means sending a light pulse to the target, measuring the distance it travels and the time it takes to return), and **gyrometers**, which measure the rotation speeds of mechanical devices. Their uses in medicine range from the destruction of tumours, polyps, angiomas and fibromas, to the treatment of caries and to ophthalmic microsurgery (see p. 167).

Metallic insulating paint
Mott, 1949; Emsey and Edwards, 1987

Ever since Lavoisier, a metal has been defined as a body which has a strong thermal and electrical conductivity. This being so, the metals are distinguished by their different atomic structures. In 1949 the Englishman Neville Mott postulated that in certain conditions of temperature and pressure, a metal could be an insulator rather than a conductor. This hypothesis was in fact a consequence of a discovery made in the 20th century — that under certain conditions (low temperatures), all elements behave like metals. In fact, in the second half of the 20th century, organic materials such as **polyacetylene** have been synthesized and are excellent electrical conductors, as well as inorganic materials such as KCP ($K_2Pt(CN)_4BrO_3.3H_2O$), which also has the same properties.

In 1987 the Britons John Emsley and Peter Edwards furthered the understanding of electrical conductivity by formulating the hypothesis that this conductivity depends on the necessary level of atoms to ensure the passage of the electrons (10 000 atoms are needed to constitute a molecule of silver, for example). One of the consequences of this property is to have enabled the American army to make paints containing metallic microparticles, which are about as small as a nanometre (a thousand millionth of a metre) and which, being insulators, would enable the aeroplanes to be 'invisible' to radar.

Metallic xenon

Ruoff, Cornell University, 1979

Xenon in metallic form was obtained in 1979 by Dr Arthur L. Ruoff and his team from the American Cornell University, who subjected xenon, a **noble gas**, to a pressure of 320 kilobars, which is 320 000 times atmospheric pressure. This laboratory feat was to open the way to the production of **metallic hydrogen**, which is used particularly in research on **superconductivity**.

Quanta

Planck, 1900

The invention of the concept of quanta (the plural of quantum) is one of the most important in modern science. Hardly a single area exists which does not have traces of its influence. The invention was originally in the field of **thermodynamics**; it consists of a culmination of theories put forward since the 1850s.

In 1859, the German Robert Kirchhoff proposed his **law of radiation** according to which the power of an emitting source of radiation depends on the absorption capacity of the medium. Then the Austrian Josef Stefan established the relation which exists between a **black body** radiator and the fourth power of its temperature (T^4) (a black body is an abstract entity which completely absorbs all wavelengths of radiation). It was from this proposition that his countryman Ludwig Eduard Boltzmann conceived the mathematical basis of the relation which has since been known as the **Stefan–Boltzmann law**. Applying himself to the phenomena ruled by this law, the German Max Planck observed the discontinuity of the emission of actual radiation from bodies; according to him, these emissions come in 'packets' or 'quanta', or in everyday language, in 'lumps'. It was in this way that Planck formulated the Quantum Theory in 1900.

This concept caused considerable fascination among all the physicists of the age, and the possibility of studying radiation in the form of quanta was revolutionary in a world where until then the concepts of only two categories of objects were used, **corpuscles** and **waves**. Corpuscles were entities occupying a defined and restricted area, also having at any moment a determined position and velocity. As for the dynamics, they were determined by their properties of energy, motion and velocity. Waves were continuous phenomena which were non-localized and occupied the whole of space. All wave phenomena could therefore be considered as a superposition of periodic waves in time and space, characterized by their wavelength.

Several observations had already indicated that there were gaps in these exclusive definitions of physical phenomena. Indeed, in 1817 the Englishman Thomas Young had shown in a celebrated experiment that photons can behave like corpuscles and waves at the same time. While actually studying the **interference patterns**, Young had postulated that these particles propagate not longitudinally, in the apparent direction of their movement, but perpendicular to it. This brilliant intuition explained the phenomenon of polarization which had until then been incomprehensible, but it was immensely inconvenient from the point of view of dogma, for it corresponded neither to the general conception of corpuscles, according to the theory of Isaac Newton, nor to that of waves, according to the theory of the Dutchman Christiaan Huygens.

Repeated with a beam of single photons and so (supposedly) not susceptible to diffraction, Young's experiment was going to prove crucial for the understanding of quanta: a single beam of photons, transmitted towards a partition with two holes in it, produces interference patterns on a screen, just like trains of waves; logically, this would only happen if the photon passed through two holes at once, which is precisely impossible. In fact, this photon behaves like a particle and a wave at once, which would be impossible to conceive without quanta. The same experiment, done with electrons, arrived at the same conclusion: the photons and the electrons are therefore neither waves, nor classic corpuscles; in order to define them one must agree beforehand that the notions of wave and corpuscle are merely approximations and so return to the concept of the **quantum particle**. Although it is impossible to represent them in pictures, their definition can be approached through Planck's suggestion that they are packets of energy which are propagated following laws of **probability amplitudes**. It is this which allows one to begin to understand that these particles, whether photons or electrons, will not strike the screen situated behind the two

holes described above as individual particles, but as amplitudes.

Two immediate consequences of the application of quanta to mechanics remain to be noted. It has allowed the development of a fundamental tool, **quantum mechanics**: all corpuscles in fact, and not only photons and electrons, define themselves as quantum corpuscles. Also the notion of **intensity** in classic wave mechanics is replaced by that of **probability density**, which is specified as the square of the amplitude of the wave.

A third consequence, the effects of which have been felt since the 1920s, is that the description of a general physical system can be done with exclusive use of mathematics. This implies a representation of the world according to mathematical formalism. Since then, the imaginary description of the physical universe as a system composed of distinct particles has become obsolete. Quantum mechanics is actually based on the use of the mathematical tool.

The fourth and final consequence has philosophical resonances: the determinism defended by scientists since Laplace and upheld ardently by Albert Einstein despite the facts, is no longer acceptable; it is

The enormous role played by Max Planck in the evolution of contemporary exact sciences was nearly seriously compromised by World War II during which he found himself suspected by the authorities of the Third Reich, not only because of the ties of friendship which he had with scientists considered as enemies (of which Einstein himself was one) but also because his own son had participated in the assassination attempt on Hitler in July 1944, and for this reason had been executed. After the war, Planck had the supreme satisfaction of seeing the ancient prestigious research institutes of Kaiser-Wilhelm renamed in his honour.

One of the most outstanding chapters in the history of contemporary physics is Albert Einstein's opposition to quantum mechanics. Having recognised its brilliant significance at the beginning of the century, Einstein nevertheless rejected this tremendous conceptual tool. What bothered him most in Planck's theory was its probabilist aspect.

Einstein, in fact, was a determinist, as is attested by his two well-known sayings: 'God does not play dice' and 'God is unpredictable, but not malicious'. In 1910 he endeavoured to cut short the inevitable progress of quantum mechanics by publishing his project on the unified field of general relativity; he realized immediately that his conclusions were premature. At the end of the 1980s nothing had yet approached, even from afar, the culmination of the Einsteinian proposal of **unified field** as Einstein had meant to do, that is, in the integral abstraction of quantum mechanics. From the 1920s onwards the great scientist's objection was to cause distress among several of his most illustrious peers, such as Max Born, who deplored the loss of the 'flag-carrier' who had been Einstein. During his last working years at the Institute for Advanced Studies at Princeton, the disillusioned Einstein endeavoured, but in vain, to fight against the supremacy of quantum mechanics.

replaced by a **statistical probabilism**. From electronics to optics and from molecular physics to astronomy, quanta have offered a huge contribution of which the riches have not yet, at the end of the 20th century, been entirely explored.

All the same, it is right to note that Max Planck himself would never support quantum probabilism, which developed essentially under the impetus of the German Werner Heisenberg, and of the Dane Niels Bohr and, in general, of the disciples of the so-called School of Copenhagen. One can assume that his classic training would explain his repugnance to what he considered to be a breakaway theory. But one might also suppose that the correspondence, which is more or less correct, between the macroscopic phenomena, such as they then knew, and classical mechanics, justified his refusal.

At this moment there are still macroscopic phenomena which it does not seem possible to explain entirely by classical mechanics, particularly in astrophysics. Besides, quantum probabilism seems to appear in certain phenomena which have only been discovered recently, such as **lasers** and **superconductivity**.

It must be underlined that the importance of this conceptual debate lies especially in the value of scientific measurements and that some scientific schools of thought still exist which refuse to allow quantum probabilism, which they hold ultimately as an epistemological artifice.

Soda (extraction of)

Solvay, 1865; Mond, c.1870

The extraction of soda has been carried out, after a fashion, since the end of the 18th century, using the Leblanc and Davy processes (see *Great Inventions Through History*). But in 1865 the Belgian, Ernest Solvay, built a factory in Belgium in which he used an entirely revolutionary extraction process. Strictly speaking, Solvay did not invent it in the sense of being its sole originator; the merit for what is called the Solvay process should undoubtedly be shared by the Frenchman Augustin Fresnel, who was the first to experiment with it and may be also the first to conceive it. The idea is very simple: **ammonia** and **carbon dioxide** react with sodium chloride; the carbon dioxide enters into a reaction with the sodium to make **sodium bicarbonate**, and the ammonia is captured by the chlorine to make **ammonium chloride**. The difference in solubility between the two salts allows the collection of the former which separates by precipitation. The separation of the sodium and chlorine is achieved, the sodium bicarbonate can be used, and only has to be heated to eliminate the carbon dioxide; it is then possible to salvage the ammonia by using lime to break down the ammonium chloride. In theory at the end of the operation **sodium carbonate** is produced.

In reality, the operation is much more complex. Fresnel, who had used solid sea salt, lost the ammonia and did not obtain any result. The German Heinrich August von Vogel demonstrated in 1822 that **brine should be used**. More than ten researchers met with failure in their attempts to find a process where neither salt nor ammonia were lost. Solvay's credit lay in his design of the plant which would ensure optimal productivity. His idea was to run the saturated solution of sodium chloride and ammonia across perforated plates which were tiered from top to bottom with metallic columns, at the base of which carbon dioxide was introduced. The ammonium chloride was treated according to plan with lime and heated so carefully that the loss of ammonia was no greater than 1%, which was nearly 30 times less than that achieved by his predecessors. Those who upheld the Leblanc process (see *Great Inventions Through History*) tried to resist the Solvay process, but the latter, being quicker and more efficient, eventually took over. This was solely due to the fact that it would provide soda at considerably more advantageous prices. The Solvay process was modified during the 1870s by the German Ludwig Mond.

It is possible by the electrolysis (passing of an electric current through a solution) of dissolved salt to obtain **sodium hydroxide**,

There is a startling contrast between the destinies of the two men who contributed the most to the industrial exploitation of soda: Nicolas Leblanc, ruined after the French Revolution, committed suicide; Ernest Solvay, who was wise enough never to give away the licence to his process, became enormously rich (he was also a great philanthropist).

usually known by the name of caustic soda. In many cases it can replace other forms of soda.

The application of the Davy process by electrolysis, was of course not possible during Leblanc's time, since it required considerable electricity consumption and power stations did not then exist.

Nowadays soda is used in the fabrication of paper, cellophane, rayon, glass, soap, fertilizers, porcelain, explosives, matches and dyes. **Sodium peroxide** is a bleaching agent. **Bicarbonate of soda** is a product used in pharmacy and bakery. **Sodium thiosulphate** is used for fixing negatives and photographic prints. **Sodium sulphide** comes into a great deal of processes in industrial chemistry.

Communication, Culture and the Media

The huge advances made since the middle of the 20th century in the basic knowledge of electronics, electromagnetism, optics, and the growing mastery of technology have released a near-explosive proliferation of different modes of communication. The spread of radio communications, which had been made possible at the beginning of the century by the birth of the radio, was vastly stimulated by the invention of the transistor. The stunning improvements in techniques of sound and image recording enabled the growth of the exchange of information to what some people thought was saturation point. The development of the exchange of symbols has constituted one of the major features of the evolution of the world since 1850.

Such rapidity in this exchange has inevitably altered the very content of information and its style, that is, the culture itself. It is the insoluble riddle of the chicken and the egg — is it the technological progress which has incited, for example, the spread of jazz and its ensuing developments leading up to rock and roll, or are these musical fashions in themselves the expression of the 'spirit of the time', the *Zeitgeist* so important in German philosophy? Similarly, the 'science fiction' created by Jules Verne has served as a kind of playing field for the scientific imagination. His prescience is famous: the point at which Jules Verne dispatched his famous shell *De la Terre à la Lune* was exactly the same as the departure of the first American space mission. And, in 1936, Hugo Gernsback broached the unthinkable, that is, the military exploitation of atomic energy. At that time no scientist really believed it possible. Five years later the Manhattan Project was started! Thus we must not separate culture and science; science is a form of culture. We should allow as much space to jazz, a real 'invention', as to the radio which first enabled its transmission.

Automatic telephone exchange

Strowger, 1889

The first telephone exchanges were staffed by operators who connected the calls by means of a **plug and socket** arrangement. Alman B. Strowger, a Kansas undertaker, invented an automatic system; it was installed in La Poste, Indiana in 1892. The switching system he had devised was developed by engineers at the Automatic Electric Co and was to be used throughout the world for the next 70 years, under the name of the Strowger system. A system of **electrical pulses**, produced originally by means of push buttons, later by a dial, caused a contact to rise and rotate to make the desired connection. During the 1960s **electronic exchanges** were developed. Improvements in the storage and transmission of information by digital networks has enabled simultaneous data and voice calls to take place, and many convenient services can be provided to the customer by the telephone companies.

Car radio

Blaupunkt, 1932

The radio was still young when the German designer Blaupunkt thought of making a model which would work from a car battery and which would have an aerial (the first car to be equipped in this way was an American Studebaker).

Cartoon

Blackton, 1906

The cartoon constitutes a kind of invention in stages. It can in fact be suggested that the precursor was the **Javanese shadow theatre** in which the shadows of cut-out and jointed silhouettes are projected on an illuminated screen, which dates at least to the 17th century; nevertheless, the **phenakistoscope** is its direct ancestor. Invented in 1831 by the Frenchman Joseph Plateau, it was composed of two discs mounted on an axle. The top disc, which was fixed, was pierced with a single slit on the edge. The lower disc had pictures on its circumference which represented different phases of a broken-down movement; this one could move. By turning the latter disc, an illusion of the movement pieced together was created as the images under the slit of the first disc rapidly flashed by, by virtue of the principle of **persistence of vision** which the cinema and television would exploit a century later. The procedure was perfected in 1834 by the Englishman William George Horner, who introduced it to the United States in 1867.

It is possible that it was the new version of the phenakistoscope which gave the idea to the American J. Stuart Blackton of making drawn cinema films which would unravel vertically on the celluloid ribbon. Two years after Blackton had designed the first known cartoon for Vitagraph — a series of simple faces — the Frenchman Emile Courtet, called Emile Cohl, a professional humorist, made the first French

Plateau's phenakistoscope, made by Chevalier in 1831. The picture shows the image in reverse. This 'toy' is used in front of a mirror, which reflects the surface shown, providing that one rotates its and observes the phases of the movement through the slits, which is here a game of leap-frog. This was undoubtedly the first optical device to exploit the effect of persistence of vision.

cartoon for Gaumont, *Fantasmagorie*.

It was, however, in the United States at the end of the 1910s that a standard method of making cartoons was found. This made use of the successive inventions of the pioneers Bray, Carlson, Terry and Fleischer. Walt Disney only produced his first cartoon in 1928, the famous *Steamboat Willie*, where the character Mickey Mouse made his first appearance.

The technique of making a cartoon is much more complex than that of an ordinary film, because the production must exactly co-ordinate the characters' movements with the scenery, which stays fixed; only the characters move. First of all a series of sketches is drawn, then photographed and shown in a rapid projector called a **moviola**. The corrections are noted before being given to a team who this time draw them onto pages of transparent plastic and then colour them. Each drawing is then photographed separately then transferred onto film. About two weeks is needed to photograph the 45 000 images on a 300-m film which lasts for about 15 minutes.

Collotype

Poitevin, 1855

Collotype is a planographic printing process, producing a print closer to a photographic print than any other printing process. This is achieved by the complete absence of screen which provides a softness and graduation of tone.

There are three basic principles: the hardening effect of light upon bichromated colloids; the fact that unadulterated gelatine will absorb eight times its own volume of water; and the natural revulsion of grease and water. The printing plate comprises a finely grained glass sheet of approximately 1cm thick, which is coated with bichromated gelatine, which is then exposed to light through a continuous tone negative. A virtual positive is obtained, by the light areas of the negative, corresponding to the shadows on the original document being hardened by the bichromate exposed to the light; the gelatin is then washed to remove excess bichromate, and the hardened gelatin absorbs the water. An oil-based ink is then applied which will be easily accepted by the hardened parts and repelled by the water-filled parts, with the intermediary areas absorbing the amount of ink according to their degree of hardening. This process was invented by the Frenchman Alphonse Poitevin in 1855, after the discovery of the effects of light on bichromated colloids.

> Collotype is a reliable process but delicate to use; it is generally reserved for short print runs.

Colour photography

Maxwell, 1861; Ducos de Hauron, Cros, 1869; Vogel, 1873

Ever since Niepce had discovered the principle of photography in 1826 (a principle touched on by the Englishman Josiah Wedgewood in 1902), he admitted his regret to his brother that he was unable to capture colour; for many years attempts were made to produce emulsions capable of holding colour. Since Newton it was widely known that white light is composed of **seven colours of the spectrum**, but no one suspected that the **emulsions** responsible reacted only to certain **wavelengths** of the bands of the spectrum, to blue and violet in particular, which explains the curiously flat white skies of the first landscape photographs. The regret expressed by Niepce was widely shared, and the research was intense, but there were no satisfactory results and numerous charlatans. One had to make do with painting the photos in watercolour.

The first to conceive of a photographic technique for the reproduction of colour was the celebrated Scottish physicist James Clerk Maxwell, in 1861. Maxwell in fact had the idea of taking successive photographs of a multicoloured object, in his case a tartan rosette, through three **filters**: the first time through a blue filter, a second time through a green filter, and the third time through a red one. The negatives obtained on the three glass plates were then projected simultaneously but in focus onto a screen. Each was illuminated by a separate light source, in front of which there was a filter of the same colour as that used when the photograph was taken. It was a remarkably laborious process but it was very important in being the first to have used **colour separation** for the reconstitution of a coloured object.

After a few years, in 1869, at the same time and without knowing each other, the Frenchmen Louis Ducos de Hauron and Chares Cros found a practical solution. In a small treatise called *Colours in photography, solution to the problem*, Ducos de Hauron explained that colours in photography could be realized using the **subtractive process**. Like Maxwell he proposed to take three negatives through **primary colour filters** — red (or magenta), blue-green (or cyan) and yellow. The positives were taken independently of each plate and transferred onto three gelatin plates composed of coloured pigments, reds for the plate taken through the magenta filter, blue and yellow for the two others. All that remained was to superimpose the three gelatins and a colour image was obtained. Or rather, one ought to be obtained, for he had not reckoned on a fundamental chemical problem: the emulsions did not react as he had hoped. In fact, the weak sensitivity of the emulsion to red meant that Ducos de Hauron had to hold the image taken through the red filter for much longer which caused the reds to begin to turn brown. The same happened with the green which tended towards yellow and for the blues which took the least amount of time to fix.

Cros himself never attempted his solution in practice but in his memoirs which he requested should be opened only after his death he nevertheless describes the process in a very exact way.

It was only in 1873 that the German Hermann Vogel, a chemist, approached the problem from the beginning and improved the sensitivity to green of the **collodion** plates by soaking them beforehand in a bath of dye based on **aniline**. By trial and error the sensitivity of the emulsions to other colours was gradually improved, the last one, red, being finally obtained only in the first years of the 20th century.

Compact disc
Philips and Sony, 1979

The Dutch company Philips and the Japanese company Sony worked together under a joint licensing agreement to develop the compact disc, to enable the true potential of **digital recording** to be enjoyed in the mass market. A plastic disc 120mm in diameter can hold a single side of over an hour of digitally encoded sound recording, stored as a succession of pits and plateaux in tracks. The disc is coated with a reflective material (usually aluminium) which either scatters or reflects back in to the photoelectric detector a laser beam used to 'read' the encoded sound when the disc is rotated at a high constant linear speed. The advantage of the compact disc over the long-playing record is its freedom from surface blemishes, so that it can approach perfection in sound reproduction. The invention of the compact disc launched on the market in 1982, was the direct result of research on the videodisc, also invented by Philips (see **Videodisc**)

Digital restoration of recordings
Sonic Solutions, National Sound Archive, 1987

Old sound recordings nearly always suffer from 'scratches' caused by distortions of the medium and background noise due to flaws in old recordings. This distortion can disfigure the original sound to the point where it is no longer recognizable. Since the 1960s techniques of **analogue** filtering have begun to be developed. Their advantages are recognized but they do involve a certain degree of **modification of the original sound**. In 1987 the firm Sonic Solutions in San Francisco designed a **digital** filtering system. It consists of recording the original sound on a hard disc of a capacity of 1400 megabits, which represents two hours of recording, then automatically eliminating the scratches using a computer which locates the interferences and suppresses them. Then the gap in sound is filled by a sound of the same frequency as that of the note which has been covered. The background noise is similarly erased on the basis of tracks which carry no recorded sound, however short they might be. This erasure takes place on the basis of dissecting the frequency range into 2000 distinct bands, which are all measured and which therefore indicate all the frequencies to be eliminated from the whole of the initial recording. Working in realtime, the computer would be forced to do some 53 million operations per second; so it works on a different time scale and takes 8–10 hours to restore an hour-long recording. This system, called NoNoise, is the rival of another which was designed independently in Great Britain, also in 1987, called CEDAR (Computer-enhanced Digital Audio Restoration). It was invented by the National Sound Archive and works on a different principle: it requires two versions of the recording to be restored and automatically compares the two recordings which are played simultaneously; then it selects the most satisfying sound from each of them.

> The NoNoise system has enabled the restoration of music records from 1928 where the composers Maurice Ravel and Serge Prokofiev played their own works on the piano or themselves conducted the orchestras which were performing their works.

Dissonance
Wagner, 1865

Until the beginning of the 19th century composers used the **diatonic** or **non-chromatic** basis of harmony in their work. Some musicians, such as Mozart in his quartet *The Dissonances*, Franz Schubert in his *Quintet in D major* and Carl Maria von Weber in his opera *Der Freischütz* had occasionally contravened the common rules of harmony. Schubert had thus overtly used **chords of diminished sevenths** which Weber in turn had also used, though for him their purpose was to intensify the harmony.

Richard Wagner was the first, in *Tristan and Isolde* in 1865, to use methods which purposefully undermined the diatonic system. These involved **ambiguous tonalities**, such as by making an augmented sixth resolve in an augmented seventh, or by placing secondary dominants in series. Distorted and unstable chords, which highlight the chromatism and drown the traditional functional structures, were also used. The institution of dissonances would later lead to **dodecaphonic** or twelve-note and **serial music**.

Esperanto
Zamenhof, 1887

It was in 1887 that the Polish optician Ludwig Zamenhof thought out a simplified universal language, which he named Esperanto. Constructed using words of European Romance origin, Zamenhof's linguistic system is simple, for the **spelling** is phonetic. All the words are spelt as they are pronounced and have specific suffixes for the nouns (o) and the adjectives (a). It includes neither gender nor verb declension and makes use of many compound words. The political upheavals which tore Europe apart in 1914 prevented Esperanto from rising to its wished-for status. With the politico-cultural considerations and demands intensifying the linguistic sense of identity, very little place was given to a universal language. Nevertheless there are now more than 100 000 people in the world who use Esperanto.

Facsimile machine (Fax)

Caselli, 1855; Meyer and D'Arlincourt, 1872; Korn, 1907; Belin, 1925; R.C.A. Western Union, A.T.T., 1920–88

The transmission of optical signals via electrical signals, is an invention with the longest and most complex of developments. The first to have the idea without ever approaching the actual invention was the famous English scientist Sir Humphrey Davy, who around 1826 managed to decipher signals at a distance by the breaking down (via electricity) of **potassium iodide**. About 20 years later his colleague Sir Charles Wheatstone was even nearer to completing the invention with his **needle telegraph**, but he obtained practical results not much more convincing than those of the Scot Alexander Bain in 1842, who was a clockmaker by trade: Wheatstone's machine, like that of Bain, only operated on graphic signs from a relatively short distance and in rather a cursory way. After various improvements including those of the Englishman Backwell and the Frenchman Pouget-Maisonneuve, the following stage was the priest Giovanni Caselli's **pantelegraph.** This was a relatively complicated system: an iron point crossed by a current was used to write onto a paper impregnated with a solution of **potassium cyanate** which it decomposed, leaving a blue mark on the paper. Despite the difficulties in synchronizing the transmitting needle and the receiving needle, the Caselli system was installed between Paris, Amiens and Marseille in 1856.

The process was improved in 1872 by the Frenchmen Meyer and D'Arlincourt. In 1907 the German Arthur Korn carried to completion an invention he had made in Paris in 1903: that of **telephotography**. Also in the year 1907 the first transmission of a photograph via electric wires took place between Munich and Berlin. It was a photograph of President Fallières, and travelled about a thousand kilometres in this way. Korn's idea comes down to the reconstruction of the image from a broken-down form; it was the prelude to the major invention of Edouard Belin, which is much quicker and more reliable and which, above all, is automatic, while the previous transmissions were manual: the image to be reproduced is fixed to a cylinder and scanned, slice by slice, by a powerful light beam; a **photoelectric cell** placed in the line of the beam transforms the light impulses into electric impulses of a corresponding intensity. This invention was the **belinograph**, of which the principle, that is, the scanning of the image and the transformation of light signals into electric signals, preceded the first really meaningful research into television (see p. 61) by two years. This principle has not changed to this day. The equipment was improved through the impetus of American firms such as RCA, Western Union and the American Telephone & Telegraph Co., Japanese firms such as Sharp, Canon, Ricoh and Toshiba, and British firms such as Rank Xerox; it culminated with the production of portable fax machines linked to the telephone, which have revolutionized communication.

Gramophone

Scott, 1856; Edison, 1877; Cros, 1877; Bell and Tainter, 1886; Berliner, 1888

Few inventions have been made in so many versions and have been the subject of so much dispute as the gramophone. The most common attributions are to either the American Thomas Alva Edison, or to the Frenchman Charles Cros or to both of them at once. In 1856 Léon Scott de Martinville, an American of French origin, invented a device consisting of a vibrating membrane transmitting its vibrations to a tracer stylus and a turning lampblack-covered cylinder; it was with this that the first graphic recordings of speech were made. This device did not allow the mechanical reproduction of sound but it was well enough known and used in acoustics laboratories for it to be remembered.

The main difference in the apparatus patented by Edison on 12 August 1877 is the steel cylinder, which is capable of the retransmission of the recorded sounds. As for Cros, in April 1877 he had published an article in which he proposed to transfer by photo-engraving on a **steel cylinder** the grooves previously recorded on a **glass cylinder** covered with lampblack. On 10 October 1877, Cros asked the Academy of Science to officially record his article, which had already been submitted in a sealed envelope in April. Indeed, he was becoming anxious about the repercussions of the research which Edison was pursuing in the same direction in America, in his laboratory in Menlo Park. Apparently, Cros's theoretical invention would have preceded Edison's realization of the same thing by only four months. However the matter is more obscure than this because recent studies (1977) reveal that the brief outline for manufacture given by Edison to his colleague John Kreusi dated 22 August 1877 was only actually made between 4 and 6 December, three months later. It was on 6 December that Edison spoke the famous phrases, firstly 'What God hath wrought', then the first words of the well-known nursery rhyme *Mary had a little lamb* in front

of a revolving cylinder covered with a thin sheet of tinfoil. When Edison took the needle back to the beginning he heard the weak but identifiable reproduction of his voice.

In 1886, the Americans Chichester Bell (cousin of the inventor of the telephone) and Charles Sumner Tainter patented an improved version of the phonograph (the word was used by both Cros and Edison) called the gramophone. In 1888 the German Emil Berliner designed an even better machine, the **gramophone**, and, in particular, completed the flat record.

From 1880 onwards Edison was occupied with the design of the **turntable**, an electric version of the phonograph; the inventor had even made a battery phonograph but it was found to be too sensitive to variations in voltage and to the weight of the records (then relatively heavy). Electrification actually appeared during the 1930s when the manufacturers produced the combined radio-phonographs.

These instruments used **induction motors** and some of them even used **synchronous motors**, similar to those in clocks but without a starting handle. It was only during the 1950s that phonographs ceased to be made with handles, as a result of the progress made in the mass production of precision instruments (the electric turntables required among other things, to be mounted on shock absorbers in order to dampen the vibrations). The commercialization of instruments with several loudspeakers connected to amplifiers meant that the turntables had become **record players**.

At first the recording cylinder was rotated by hand which very much affected the sound; but in 1878 the problem was solved by the addition of an electric motor. In Edison's laboratory there was even a **steam-powered phonograph**.

Edison in 1897, *39 years of age, showing a recording cylinder covered with tinfoil. Above left: the phonograph made by Edison in 1878.*

Jazz

Collective, c.1850

Although there is no unanimous musicological definition of jazz, it is considered to be at once a major cultural invention and itself the expression of a more or less determined musicological field. Its origins are obvious but anonymous because they are collective. Jazz is the product of the culture shock experienced by the Black slaves brought to the United States, and, inversely, the reaction of the United States to these totally new musical styles. One point unites historical opinion: jazz as we know it would not have existed had the Blacks never come to America.

The ground offered for the genesis of Black music which was exceptionally favourable to the development of jazz was **communal singing**; it had been very much developed during the 19th century in the United States and was just as much lay as religious. The essentially farming communities which had grown up in the New World were accustomed to have gatherings where they would often sing **folk songs** from their homeland (Ireland, Germany, England, etc.). Similarly, the slaves in their enforced communities would also sing together, but following styles, scales and harmonies totally unfamiliar to Westerners, dictated much more by intuition than by rigorous musical knowledge. The main characteristic of this music was the communication of emotion.

In Black music which is essentially from West Africa, the **mediant** and **dominant chords** were transcribed in **minor** or in **relative** key, which obviously surprised Western ears and produced a rather melancholy sound, which led to these chords being known as 'blue notes' since blue is a twilight colour, and the Blacks sang their 'blues' at dusk. It was undoubtedly in this way that the first form of jazz, the **blues**, was born.

Long before the 1850s, groups of travelling musicians were set up in farming States in order to entertain the communities who had hardly any entertainments. They performed their own compositions based on folk songs but which were already showing signs of Black music. From the beginning therefore, the American **folk song** would show traces of Black influence which set it apart from European folk singing. We note in passing that this tendency was to lead to the production of musicals such as Jerome Kern's famous *Show Boat* or *Porgy and Bess* by George Gershwin, which were hybrids of White music and Black music but nevertheless characterized a profoundly original and enduring musical form.

In the same way, the Black community singing was affected by the White song and resulted in an entirely new form which was the **negro spiritual**. This was essentially inspired by religion but also performed at non-religious occasions.

The new style won over instrumental music by means of the saloon bars and shows and through attractions as diverse as the piano accompaniments in the first cinemas and interludes in fairs. According to numerous musicologists the gestation of jazz took place during the last third of the 19th century, about which one cannot be more precise. Before the actual instrumental jazz appeared there was a particular form called **ragtime** which was composed of pieces of two or three distinct sixteen-bar themes. Being strongly **syncopated**, ragtime was to transmit to jazz a distinct tendency for abrupt statement.

From the beginning of the 20th century jazz pieces such as *Memphis Blues* (1911) and *Saint Louis Blues* (1914) composed by W. C. Hardy had achieved a notoriety which without being national was already crossing regional frontiers. These were instrumental blues composed of twelve-bar sequences and only contained three critical points of harmony. The performer could therefore improvise freely with his variations on the melody according to his personal style. At first these pieces were simply memorized and often varied a great deal from one musician to another. Subsequently, **improvisation** was to make such an impression on jazz that it even became one of its major features.

During the first two decades of the 20th

century jazz was to produce quite a school of instrumental creation centred around the town of New Orleans. It is said that between 1905 and 1915 there was a 'second invention' of jazz based on the use of the **cornet** (the ancestor of the trumpet) and the **clarinet**, with the **trombone** as 'harmonic bass'. The most notable artists here were Freddie Keppard, Jerry Roll Morton, Sidney Bechet, Louis Armstrong and Joe 'King' Oliver. The success of these bands is partly due to the fact that despite the small number of musicians (five were easily enough to give a concert), they

produced an impressive volume of sound; it did not cost much therefore to liven up marriage ceremonies as well as funerals or street festivities.

The incorporation of jazz into the American culture of the time was helped by the recordings on discs, on perforated bands for the mechanical piano, and then the radio. In 1917 the first great White jazz orchestra, the famous Original Dixieland Jazz Band made its debut in New York. Shortly after World War I jazz became international and caused both fury and scandal in Europe.

The importance of the influence of the collective invention of jazz can be measured in the repercussions which resulted from it in the arts and the cultures of the age.

The originality, particularly of the syncopated rhythm and the violent colour of the brass was to influence not only the 'cultured' music (Shostakovich, Stravinsky), but also painting, literature and decorative art.

During the 1960s the most recent derivative from jazz (the first being the fox-trot, the boogie-woogie and the be-bop), **rock and roll** (itself derived from boogie-woogie) was going to give way to a real international social phenomenon which transcended the different politics between the West and the Communist block, for Soviet rock groups also appeared. Though defined at the outset as 'sub-culture' the whole of the musical domain dominated

by rock and roll as well as its infiltrations into different modes of expression such as journalism, radio and television and associated activities, not forgetting fashion, was to promote an international popular culture. Following the Woodstock Festival which was attended by the previously unheard of number of 400 000 people, and also affected by 'monster' concerts (Altamount, Isle of Wight, etc.), rock adopted a particular ideological hue. This was politically pacifist, in favour of all kinds of minority groups and of more or less leftist tendency, the historical role of which cannot yet be estimated.

It must be noted that what is stated above does not claim to constitute a complete history of jazz, which involved and still involves numerous variations, such as **hard-bop**, **jazz-rock** and **funk**.

Linotype®

Mergenthaler, 1886

The rapid movement of the **popular press** owes much to the invention of the Linotype®, type-casting machine. Prior to this invention type was hand set which was slow and laborious. The Linotype® was patented by a German-born watchmaker, Ottmar Mergenthaler of Baltimore, in 1885 and comprises four main sections: the keyboard which is operated to release matrices from the magazine; the magazine which contains brass matrices of each individual letter; the casting mechanism

which brings the molten metal in contact with the assembled matrices to cast and deliver a solid line of type known as a slug; the distribution mechanism which removes the line of matrices after casting and returns them to their correct position in the magazine ready for further use.

Justification of the assembled matrices, prior to casting, is achieved by wedged shaped spacebands, which are placed beween the assembled words, being forced up to make the line of matrices fill the measure.

Long-playing record
Harrison and Frederick, 1931; Hunt, Pierce and Lewis, 1933; Goldmark, 1948

The arrival of the long-playing record in 1948 followed much technical and theoretical experimentation. Ever since the first records in 1901 and 1903, which were cylindrical at first, then flat, the material used had been rather hard with deep grooves. The needle was fixed firmly in the pick-up cartridge and shaped with an angled shaft, the point of which rested in the bottom of the groove. The cartridge had to be heavy enough for the needle to follow the shape of the groove exactly. With the pressure of the needle on the material the surface noises tended to pick up the recorded sound. In 1931 the Americans H. C. Harrison and H. A. Frederick proved that the sound quality could be improved by using a softer recording material and a lighter cartridge. This discovery entailed an unwelcome commercial upheaval, for all the equipment for recording and reproducing sound, including the discs, had then to be replaced. Indeed, there were already tens of millions of records on the market which were likely to be made obsolete, firstly because people were unwilling to spend money on this type of commodity due to the crisis resulting from the Wall Street Crash in 1929, and also because the radio was competing more and more strongly with the gramophone. For all that, the **magnetic pick-up cartridges** recently on the market were heavy too and since they were used by radio stations among others, could not just be rendered obsolete. In 1932 they were made lighter but were no longer suitable for the automatic record-changing system which was popular at the time.

In 1933 the Americans F. V. Hunt, J. A. Pierce and W. D. Lewis made progress towards the long-playing record in proving the advantages of the sound production obtained with the more open radial recording and a lighter, round-pointed needle. Henceforth, the very fine grooves were in sight, which would number some 100 in every radial centimetre, as well as a light cartridge and a much more delicate stylus which rested on the edges rather than the bottom of the groove. Although he had the benefit of these discoveries, which he improved upon, it was not until after World War II that the American Peter Goldmark (of the firm Columbia) produced the first 30-cm wide record which played at 33⅓ rpm. The playing time of one side had changed from about 5 minutes for the large 78 rpm records to 25 minutes and the sound quality was considerably improved. The 78 rpm records vanished quickly from the market.

Magnetic recording
Poulsen, 1898; O'Neill, 1927

The principle of magnetic recording involves applying electric impulses which correspond to the sounds to be recorded to a magnetizable metallic base. The sounds are then played with a magnetic cartridge which interprets the differences between the strengths of the recorded magnetic impulses as sounds. This process was invented in 1898 by the Dane Valdemar Poulsen, who presented it to the Universal Exhibition in 1900. The first magnetic recorder worked on a fast wind wire. Magnetic recording became widespread from 1927 when the American J. A. O'Neill replaced the wire with a **diamagnetic ribbon** coated with a metallic substance. The most commonly used are **ferric oxides** and **chrome dioxide** on a Mylar tape.

Photographic film

Eastman, 1889

Before the photo film was made on a reel, invented in 1889 by the American George Eastman, the **photosensitive media** had been the **sheet of albumened glass** (Abel Niepce de Saint-Victor, 1847), **albumen paper** (Evrard-Blanquart, 1848), the **collodion plate** (Gustave le Gray, 1849, inventor also of the technique of landscape photography using two negatives which gave the sky 'relief'), the **dry plates** of **bromide on gelatine** (Richard L. Maddox, 1871), and the **negative paper** (George Eastman and William H. Walker, 1884). The latter one was the first to be put into a reel; it had to be mounted on glass for prints to be produced. The ultimate step was made with the **transparent film** which would thenceforth replace the glass and be used directly for making prints.

Phototypesetting

Porzolt, 1894; Higonnet and Moyroud, 1944–54

There are many conflicting opinions on the exact circumstances of the origins of photo-typesetting.

Some sources claim that the first prototype was invented in 1894 by the Hungarian Eugene Porzolt. However, this complex machine was never exploited commercially.

The first phototypesetter to be successfully developed was developed by a French engineer named Rene A. Higonnet, who in collaboration with a colleague, Louis Moyroud, evolved a machine that projected light through a spinning disc containing a type font, onto photosensitive material. To seek financial backing Higonnet visited the US in 1946 and gained the support of William Garth of the Lithomat Corporation. After further development the first machine was demonstrated to the industry in 1949. Later Lithomat changed its name to Photon and in 1955 a series named the Photon 200 was launched.

The further development of photosetting can be broadly categorized into four generations.

At the same time that the Proton 200 Series was being developed, the major hot metal typesetting machine manufacturers were introducing phototypesetting machines and in 1950 Intertype launched their Fotosetter. (This machine was the first phototypesetter to be installed in Scotland.) Mergenthaler, the manufacturers of the Linotype hot metal typesetter, introduced the Linofilm and Monotype introduced the Monophoto. The Monophoto had the greatest success and became widely established. All of these machines were modifications of the existing hot metal machines, with the matrix modified to contain a negative form of the font character. Each character was exposed individually after being positioned alongside the preceding character.

Unlike first generation machines which were modifications of existing equipment second generation machines were designed specifically as phototypesetters. Their design incorporated electro-mechanical escapement systems with electronically controlled mirrors, lenses and gears. This generation of machine saw very important developments and the introduction of more sophisticated devices enabling the setting of lines of type instead of individual characters. They also introduced character storage on drums, discs and grids which provided a wide selection of size changes without the high capital outlay required for hot metal typesetting and first generation machines.

Throughout the 1960's we saw the development of the third generation machines. These machines were developed using electronic escapement devices instead of the electromechanical devices used earlier. By introducing cathode ray tubes it was possible to generate digitized fonts instead of using fonts generated from a matrix. This was achieved by generating on the face of the cathode ray tube and exposing it onto photographic material. This development also established very high speed typesetting.

Fourth generation machines are generally associated with the introduction of digitized laser techniques where a laser replaced the cathode ray tube to produce the image. They were developed in the early 1970s and resulted in very high speed output. Among other sophistications to be developed were front-end systems enabling editing and correction to be carried out on visual display units prior to processing through the phototypesetter. Some machines allowed merging of illustrations within made-up pages, also prior to final output. Normally such data generated by front-end systems was stored on magnetic tape which was subsequently used to drive the phototypesetter.

> Phototypesetting has continued to make some very significant developments. Factors which have contributed to this have been the introduction of Page Description Languages such as Postscript and the concurrent development of Imagesetters producing illustrated make-up pages direct from front-end systems. Currently tests are being carried on a system which will interface the front-end systems directly to a roll of 70mm film which is then, under computer control, exposed in accordance with a pre-determined imposition onto a lithographic printing plate. This eliminates the need for the camera and the assembly of the film on foils.

Pneumatic conveyor

Clark, 1853; Varley, 1858; Culley and Sabine, 1859

The system of sending mail through vacuum tubes was invented in 1853 by the Englishman C. Latimer Clark. The first one was only meant for the dispatch of telegrams between the Central and Stock Exchange and the International Telegraph Company; the cylinder in which the telegrams were enclosed, and travelled in a tube where the vacuum induced a **suction effect**, only traversed a distance of some 220 m. In 1858 the Englishman C. F. Varley made an improvement by forwarding the cylinders using **compressed air**, which allowed them to travel in two directions in the same line. The following year, the Englishmen R. S. Culley and R. Sabine organized a **pneumatic mail distribution network** in the city of London. When the system was adopted in Paris in 1868 it was altered: the layout of the tubes (400 km) became circular and the circulation took place in one direction only, either by the force of a vacuum, or compressed air.

Polaroid

Land, 1947

The **instantly developed photograph**, known by the name of the firm Polaroid which previously specialized in the manufacture of polarization filters, was announced to the press on 27 February, 1947. It received a luke-warm welcome, as the instant photograph – in black and white at first – had been described as an object which resembled a photograph. The inventor, Edwin H. Land, only achieved the colour photo in 1963, with the **Polacolor**, followed in 1972 by the more improved **SX-70 film**. The principle of these films rests on the very thin layers of chemicals, some sensitive to blue, green and red light, the others consisting of developers. On one of its edges the film is equipped with capsules of developing product containing a **reactive alkaline** and an **opacity agent** which has a base of **titanium dioxide**.

This prevents the light from penetrating into the sensitive layers during development and spoiling them; the capsules are crushed between two rubber rollers in the camera as the film is exposed to the light.

> The instant photograph was actually invented by the photographic firm Agfa in 1928 but the patent, which was never commercialized, became obsolete when Land got to know of it and perfected it.

Radio

Fessenden, 1902; Majorana, 1904; 1909–12; Fleming, 1906; Pickard and Dunwoody, 1906; De Forest, 1907; Barthélemy, 1910; Sarnoff, 1916; Black, 1930

It was in 1887 that the German Heinrich Hertz discovered the waves which were named after him at first and which are now called radio waves. They are between 6 mm and 10 km in wavelength. They carried **telegraph signals** with ease because the power could be increased, there was no need for lines, and their reception using a tuned aerial was easy. It was thought therefore that before the end of the 19th century it would be possible to use radio waves to carry electrical signals converted into sounds, such as the human voice, but the realization of the idea appeared impossible due to their very low frequency. The plan was not abandoned however.

Right at the beginning of the 20th century several technicians proposed to make use of waves of very high frequency and power, but with variable amplitude: the sound signals would modify one of the parameters of the wave, either its **frequency** or its **amplitude**. This brilliant principle is the one which is now employed in **amplitude modulation** and **frequency modulation**; at the time it made reception rather difficult, for it required considerable power transmission, without which virtually nothing could be heard. Due to the dispersion of radio waves, a 50 kilowatt transmitter could only produce a few microwatts at reception. Moreover, the modulation technique had not yet been established.

The first to have the idea of the modulation process was the American R. A. Fessenden: in 1902 he experimented with frequency modulation using an Anderson alternator which transformed mechanical movement into electrical energy with the help of the vibrations of a **microphone** cooled by water. Following this principle, Fessenden was to manage the first Christmas transmission in 1906. Transmitting from Brant Rock in Massachusetts, on a frequency of 50 kHz (kilohertz) with a power of 1 kilowatt, he put two musical tunes, a poem and a talk on the air which were heard by radio operators several hundred kilometres away. Among other attempts at frequency modulation, that of the celebrated Italian physicist Enrico Majorana must be mentioned. In 1904 he managed to obtain modulations from the displacement of air by the human voice. But the modulations were too weak and the reception much too faulty for the radio to provide an effective means of communication.

The first important stage of the development of radio came when the Englishman Ambrose Fleming, advisor to the company founded in Great Britain by Edison, had the idea of using the peculiar phenomenon discovered by the American researcher while he was endeavouring to perfect the

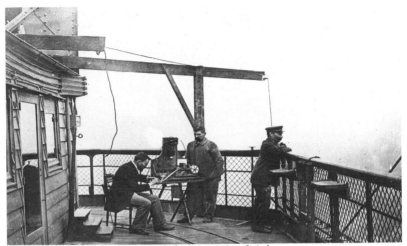

*On the top platform with the pioneers for the first transmission of wireless
telegraph between the Eiffel Tower and the Pantheon (29 July 1898).
The technician wearing a jacket is Ernest Roger; his partner at the Pantheon was Eugène Ducretet.*

*A **'bourgeois' radio installation** around 1920 in Paris. Notice the
bulky aerial and the fact that the bulbs are naked.*

invention of the light bulb: under certain conditions of vacuum and voltage a bluish glow appeared between the extremities of the filament inside the light bulb which followed a 'current' flowing in the opposite direction to the electric current. This phenomenon, the '**Edison effect**' puzzled numerous scientists and it was the work of the British physicist Sir Joseph John Thompson in particular which established its source: it was a current of electrons from the cathode (negative) to the anode (positive) as a result of heating. This transfer of electrons was defined as **thermionic cmission**.

Fleming had the idea of using this effect to receive radio waves. He put a cathode and an anode in a vacuum tube and, using a battery, heated the cathode until it was white-hot. When put in place in a radio receiver circuit, this tube had the following effect: the anode became alternately positive and negative according to the fluctuations in the received radio waves; when it was positive it attracted the electrons and when negative it repelled them, which meant that the current could only travel in one direction. This rectification which was also a system of detection and demodulation, led to the invention of the **diode** in 1905. This is used essentially for radio transmission and reception which are only possible with unidirectional and non-alternating currents.

The signals captured with the help of the diode were weak, but not for long: in 1907 the American Lee de Forest improved the diode by adding to two classical electrodes a third control electrode or grid. This functioned not only as a detector after the fashion of a diode, but also as an amplifier. This **triode** improved both the reception and the transmission.

By continued improvements the radio was brought to life. In 1910 the Frenchman Joseph Berthenod tested the first **mobile transmitter** which was mounted on an airship (then in 1914 on a Farman aeroplane). And, importantly, in 1919 the Frenchman René Barthélemy made use of the properties of diodes and triodes to complete the **mains radio** which paved the way for the age of large-scale radiotransmission since batteries and accumulators which were indispensable up until then, were no longer needed.

Until then, radio listeners built their own radio sets. The very first radio set was undoubtedly the one built by the Italian Guglielmo Marconi, using a Tesla coil, a coherer, a spark gap and an aerial. From 1906, however, it was the **crystal sets** which were the most popular. Galena is a crystal which has the property of rectifying

the current, and this constitutes the basis of the invention of the galena receiver by the Americans Pickard and Dunwoody. The galena receivers were inconvenient however, because, in order to detect a low-frequency current, a 'cat's whiskers' (sensitive point) had to be run over the crystal. Only certain parts of the crystal were sensitive; the signal received was very weak and in order to hear it, it had to be connected to a **loudspeaker**, which itself was connected to a headset. It was only possible to hear the radio under these conditions; this remained true until after the invention of the diode and the triode.

The completion of the first **diaphragm and trumpet loudspeakers** encouraged the American David Sarnoff to undertake the industrial manufacture of diode and triode radio sets, which were encased (originally in wooden boxes) so that it was possible to listen in as a group without any special equipment. This was an enormous success and it prepared the arrival of radiotransmission to a large audience.

Most of these receivers appeared in 1921. The first radio programme in France, by Radio-Paris, took place from the Eiffel Tower in December 1921. The reception was undoubtedly better than with the galena radios but there was still interference and it sounded cavernous. In 1930 this was dealt with by the American H. S. Black with the invention of **negative feedback** which consists of re-introducing into the circuit at transmission and in phase, a fraction of the outgoing signal. The invention of the transistor in 1948 by the Americans Bardeen, Brattain and Schockley (see p. 84) shortly after World War II was to give a considerable boost to the radio by encouraging the manufacture of small sets which worked from batteries and were therefore portable.

In the UK, a brief and spasmodic service was provided by the Marconi Co broadcasting twice daily from its Chelmsford works during part of 1920. The first broadcast by a professional artiste was a performance given by Dame Nellie Melba on 15 June 1920. The BBC was formed in 1922, its first programme a news bulletin broadcast on 14 November of that year.

The first regular radio programmes began on 2 November 1920, transmitted at 8 p.m. by the Westinghouse Electric and Manufacturing Company at East Pittsburgh in Pennsylvania. Its aim was to announce the results of the presidential election between the candidates Harding and Cox (it was Harding who won).

Record

Edison, 1878; Berliner, 1888

The invention of the flat gramophone record as we know it today is attributed to Emil Berliner. The idea behind it was in fact originally outlined by Thomas Edison in 1878 and Berliner is credited with having been the first to manufacture and improve it. He was also the founder of the first record moulding factory to use a master, in Hamburg. His original company was the Gramophone, later renamed Victor Talking Machine Ltd.

> The cylinder existed alongside the flat record for many years. In 1894 Charles and Emile Pathé built a very successful cylinder factory near Paris and by 1904 there were some 12 000 titles in the Pathé catalogue.

Science fiction

Verne, 1862; Robida, c.1890

Science fiction should be thought of as a major cultural invention, because it has fulfilled – and continues to fulfill – two important roles: the first is to familiarize the public with the potential uses of science and technology; the second is to serve as a testing ground for the imagination. It sometimes resembles and identifies with adventure stories, and so is particularly successful with the young. Perhaps one of the first examples was Jules Verne's *Voyage au Centre de la Terre* (Journey to the centre of the earth) (1864), but it really only began with *De la Terre à la Lune* (From the earth to the moon) and *Autour de la Lune* (Around the moon) from the same author. Unaware of the potentialities of **rockets**, which in the 20th century would enable in reality the voyages into outer space which he merely imagined, Jules Verne used a gigantic explosion to propel the interplanetary engine which was in the form of a shell. This was impossible; nevertheless he can be credited with foresight in his choice of departure point for the interplanetary engine; this was situated in Tampa Town in Florida, not far from Cape Canaveral and the Kennedy Space Center.

During the 1890s the illustrator Alfred Robida also popularized certain scientific and technological projects, displaying an incisive and often humouristic imagination. In *The 20th Century* his reference to domestic television was well before its time; this was one of his two major science fiction works, the other being *20th Century War*.

Science fiction in England was pioneered by H. G. Wells with his allegory *The Time Machine* (1895). He wrote several other science fiction titles, notably *The Island of Dr Moreau* (1896), *The War of the Worlds* (1898) and *The First Men in the Moon* (1901).

Wells, Verne and Robida often showed incredible foresight. Similarly, from 1936 onwards the American Hugo Gernsback imagined atomic energy although the scientists did not believe it possible.

> Jules Verne consulted numerous scientists before writing about spacecraft and adventures. It was in this way that he could paint an almost plausible picture of the **underwater** *Nautilus*, the mechanical hero of *20 000 Leagues Under the Sea*.

Prescience of scientific fiction: *Above, the open sea recovery of the craft launched from the Earth to the Moon in the novel of that name by Jules Verne (1865) and below, the recover of the* Apollo 8 *capsule on 27 December 1968 after ten orbits around the Moon, exactly one century and three years later.*

Stereophonic record

Waters, 1920; Cook, 1951; collective, United States, 1958

The impression of 'depth of sound' produced by the diffusion of the same recording through two sources seems to have been recognized since the birth of the record. In 1920 the American Samuel Waters patented a reasonably complicated process which had unforeseen commercial repercussions; this was the idea of playing two different pick-up cartridges over two separate grooves. A fine demonstration of this was given to the engineers of the Bell Telephone Laboratories in 1933 when a symphony concert given in Philadelphia was transmitted to Washington using two sets of loudspeakers with each one emitting the sounds of the orchestra corresponding to its position. The **stereophony** was the most effective when the two speakers reproduced different and complementary sounds according to their specific spatial positions. This is the theory that the American Emery Cook applied in quite an elementary way in making double-grooved records, with some grooves near the centre and the others near the edge, which were played by two separate cartridges. The real answer came in 1950 with the arrival of the **microgroove**, which meant cutting two edges on the surface of each groove. These would be simultaneously deciphered by one special cartridge and transmitted separately via two loudspeakers. This cartridge could also reproduce monophonic recording but the stereophonic record could only be properly played using a stereophonic cartridge.

Telephone

Reis, 1861; Gray and Bell, 1872–75

The first person to achieve the **transmission of sound from a distance** seems to have been the German Philip Reis. Ever since the Englishman Michael Faraday had demonstrated in 1831 that the vibrations of a metal could be translated into **electrical impulses**, all the necessary technological equipment existed for the making of a telephone; but it was not until 1861 that Reis got around to doing it. His invention was simple: it consisted of an **electric circuit** comprising a **metallic point** in contact with a metallic band which rested on a **membrane**; this was the **transmitter**. The **receiver** consisted of a **metallic needle** inserted in a **coil** resting on a **resonance box**. The length of the needle varied according to the electrical impulses transmitted by the metallic point of the transmitter. This followed the principle of **magnetostriction**, that is, the property of a metallic rod to change its length according to the intensity of the **magnetic field** in which it is placed, which in this case was created by the coil. Reis, who was the first to call his device a 'telephone' succeeded in transmitting speech and music in 1861, according to numerous accounts. When in a trial he questioned the patent that had been deposited by the Scottish-American Alexander Graham Bell in 1876, his claim to precedence was rejected on the grounds that his device did not work in the way he described. In fact, it did work but in a somewhat unsatisfactory way for it was rather insensitive and very loud sounds could cut the connection between the point and the magnetic band.

Bell and his compatriot Elisha Gray set to work on the problem in an indirect way: by trying to find a way of transmitting several telegraphic messages at the same time using one electric wire. Gray designed the apparatus from the electrical point of view, and Bell from the acoustic point of view, and together they came up with the

*In front of a solemn audience, Graham Bell makes the **first long distance telephone communication** in 1892, between New York and Chicago (about a thousand miles).*

principle of the **harmonic telephone**. This completed several sheets of metallic vibrators each vibrating at a determined frequency with the help of individual **electromagnets** both at transmission and reception. The simultaneous transmission and reception was achieved by connecting all the electromagnets to one single wire. Each of the inventors realized that the system could be used to transmit different frequencies of the human voice, and it was in this way that they began on the telephone. In 1874 Gray built a **steel diaphragm receiver** which was placed in front of an electromagnet, the principle of which was identical to that of contemporary receivers though he did not have a transmitter. The following year Bell replaced the steel diaphragm with a membrane and had the idea that a steel element in contact with the membrane and placed in front of an electromagnet would transform its vibrations into electric-

al impulses and transmit them back to the magnet. He attempted this but came to no conclusion. Gray on the other hand became discouraged, having designed a **mobile membrane transmitter** which had two

The invention of the telephone, of which the patent is often cited as the most profitable in the history of inventions, released a great number of lawsuits, firstly between Bell and Gray, then between the company founded by Bell and the Western Union, who had bought the patents of Gray and Edison in order to allay the success and pretensions of the Bell Company. Edison in fact had appreciably improved Bell's device by incorporating a **carbon disc** into the transmitter, which greatly increased its sensitivity. In 1893 Bell ended up winning all his lawsuits and went into history as the unequivocal inventor of the telephone.

needles, one of which was submerged in an acid bath to render it conductive. This system resulted in the variation of the electrical resistance according to the distance between the needles and thus the variation of the current in the circuit. Both men were on the right track without knowing it: if Bell's laboratory had been less noisy, the inventor may have realized that his telephone did work well although it was still not very sensitive. Gray, for his part, had not even built his transmitter, which was nevertheless more effective than Bell's diaphragm. It is generally thought that his discouragement was caused by the delusion that Bell had got there first and that the transmission of the voice from a distance really was not important anyway.

On 7 March 1876 Bell registered the patent of his invention and, three days later, after testing a type of transmitter like Gray's, he succeeded in transmitting the voice. The invention only became reliable in 1877 although Bell demonstrated it at the centenary festival of Philadelphia in 1876. It was commercialized at the beginning of 1877, when the first telephone exchange in the world was installed at Hartford, Connecticut where it served 20 lines. The first intercity connection was made in 1883 between Boston and New York.

The first exchange in the UK open to general subscribers was established in London in 1879.

The quality of telephone connections was greatly improved by the use of the **triode amplifier** which was invented in 1906 by the Austrian Robert von Lieben.

Until the middle of the 20th century the telephone was considered a secondary means of communication in most of industrial Europe. In 1950 there were 5 million exchange line connections in the UK. In 1960 the figure was about 7 million. The next 30 years saw a tremendous increase in telephone use, until there were nearly 24.5 million exchange lines by 1989.

Television

Nipkow, 1884; Braun, 1897; Wehnelt, 1898; Vichert, 1899; Mihaly, 1919; Baird, Barthélemy, de France, Chauvière, 1920; Zworykin, 1934

The principle of the television goes back to 1884, the year when the German Paul Nipkow registered the patent. It involves the conversion of a picture into a series of lines made up of points of different intensities of light and its transmission using **electromagnetic waves** of different frequency and intensity, which are themselves converted into currents also of different frequency and intensity. Inspired by the discovery of the **photoconductivity of selenium** in 1873, Nipkow's idea was based on a perforated disc which scanned the image at transmission and the screen at reception. At transmission, the perforations only allowed one light signal through at once; at reception, only one was sent as well, this signal being modulated by a **selenium amplifier**. This was the principle of **automatic sequential scanning**, which remained in force until the discovery of **electronic scanning**.

However, it was not until 1927 that televisions became reliable. This was based on the following principle: a special camera, one which can be used for filming, comprises a cathode tube and transparent screen. Behind the screen, an **electron gun** emits a beam of electrons which scans the whole image systematically; each time the beam hits the screen an electric current is produced. The beam continues point by point along very tightly packed lines. The intensity of the electric current is proportional to the luminosity of the point; a white spot is very luminous, a grey one much less so, and a black one not at all. These electrical impulses are transmitted by cable

The Nipkow disc,
an invention which
played an important part
in the development of the
television, here in its
so-called integra version.

towards the **treatment centre** of the transmitter; they constitute signals which are amplified, corrected and synchronized with the sound. The transmitter produces a wave with the help of a stable-type **oscillator**; this is called a **wave carrier**. It is modulated by the **video signal** before being released by the **aerial** to go to the **receiving stations**, and it is captured by an aerial which conveys it via a cable to the

The actual idea of television was conceived by the American George Carey in 1875, but it was hardly practical because it was proposed to have as many lines of transmission as there were spots in the image. In 1880, nevertheless, the American W. E. Sawyer and the Frenchman Maurice Leblanc, the 'creator' of Arsène Lupin, had the brilliant idea of systematic scanning which was to be patented by Nipkow.

television set. Once it is again treated by electronic circuits the video signals are extracted and injected into the cathode tube of the receiver. Here again an electron gun sprays a luminescent screen with these signals, point by point, line by line; it is a symmetrical operation to that of the filming.

This spraying takes place very quickly, recreating the whole picture in a very short time. If it were slowed down the viewer would perceive only a series of spots of varying luminous intensity; in fact, the persistence of the light signal on the retina, also known as persistence of vision, means that the luminous impression of the first line still has not disappeared when the electron gun scans the last line.

This principle was established essentially in 1881 and it was patented in Paris, London and New York. But the technical equipment needed for its application was

Felix the Cat, the famous cartoon character, as he appeared in a televized experiment in 1929 (on an image of 60 horizontal lines) and as he appeared in 1937 (on a 455 line image).

lacking. The first to appear was the **cathode tube** invented by the German Karl Braun in 1897, who approached the problem of the scanning of the beam; but all he did was to point the way forward, for his focussing was insufficient to break down the picture into distinct **spots** or **pixels**, and the beam was too weak anyway. The following year, the problem of intensity was solved by the German A. Wehnelt, inventor of the **alkaline oxide cathode**, which improved the intensity of the transmitted electrons, but the problems of focus remained. The following year again, the Frenchman Vichert invented the 'concentration electrode' which improved the focusing. Television, imperfect though it might be was possible. In 1919, the Hungarian Denes von Mihaly made a transmission of pictures of instruments and moving letters over a distance of a few kilometres and this aroused enthusiasm.

Several high-level specialists, René Barthélemy, Henri de France and Marc Chauvière in France, John Logie Baird in the United Kingdom, and Schroter in Germany endeavoured from the 1920s onwards to improve transmission and reception. In 1927, Bell Telephone organized a direct transmission between New York and Washington, based on an adaptation of Nipkow's disc which the American G. S. Carey had made in 1873: he had designed a way of projecting an image onto a screen consisting of thousands of selenium cells, each one of which was connected by an electric wire to an electric light bulb; each selenium cell under the impact of light emitted an electric current of an intensity corresponding to that of the light received, and the screen of bulbs recreated the image. The transmitter used in 1927 comprised 2500 selenium cells and the receiver screen was made up of as many neon lights.

The device would have been developed despite its weight and its cost if, in 1934, the Russian, Vladimir Kosma Zworykin, who in the meantime became a naturalized American, had not invented the **iconoscope**. Using this the image to be transmitted was projected onto a screen of photoelectric cells at the bottom of a cathode tube; each cell lost its electrons with the impact of the light, but the electron beam restored the same amount, and the induced current took on an intensity corresponding to the level of brightness of the cell and released the video signal during the scanning. This was the direct forerunner to the modern tube.

In 1928, John Logie Baird patented a colour television process which he had completed experimentally at least; it was based on the observation that our colour vision of the world could be made up from the three colours – red, green and blue. Baird thought of covering the perforations of a Nipkow disc with filters of these three colours. This is the principle which in the 1950s culminated in the **shadow-mask tube**: the screen was covered with triplets of **phosphorus pastilles** of these three colours, which the three electron guns scanned, one for the red, another for the green, the third for the blue. The pastilles or luminophores acted just like the black-and-white pixels.

The first television transmitted in France was installed on the Eiffel tower on 10 November 1935; it transmitted for a few experimental receivers, in 180 lines. The first commercial television transmitter in the world was installed in 1936 by the Radio Corporation of America on the Empire State Building in New York; it transmitted on 343 lines and changed to 525 in 1941; it had only 5 000 viewers then. Although television was a luxury item, the BBC already reached 20 000 homes in 1939.

Germany had already produced a colour television of exceptional quality in 1939 with a 1029 line definition (Telefunken), but public television was not known there until after 1953. The American standard was fixed at 525 lines, the European at 625; during the 1980s a new high-definition system was developed, but different standards were proposed by Japan (1125 lines), Europe (1250 lines) and the USA (1050 lines), leaving manufacturers unsure of which standard to follow.

Three-dimensional television

Baird, 1928; IRT, Philips, JVC, Sharp, Fisher, Tektronics, Stereographics Inc., Atari . . ., Odet and INSA, Chauvière, Guichard and CNET, 1980

Three-dimensional pictures were in fact invented in the 19th century (see *Great Inventions Through History*) developing from a principle that several firms and technicians have applied to television several decades later. This principle which is very simple in theory consists of making two images of the same object in two complementary colours, A and B, which are then shown to the viewer after they have been superimposed. The viewer is equipped with special glasses of which each lens corresponds to one of the complementary colours. The eye behind lens A will perceive only image A and not image B and vice versa, so the illusion of three dimensions is made in the brain by a trick of **perception**. In 1928 John Logie Baird had already patented a three-dimensional television system which enabled the transmission and reception of stereoscopic images (by separation). Some attempts at three-dimensional cinema based on this principle took place during the 1930s.

From 1980 onwards firms and researchers connected to television began to design an application of this principle for the small screen. In 1982 British, American, German and French television channels began to transmit black-and-white films which could be seen in three dimensions provided that the viewer was equipped with stereoscopic glasses. In 1987 the Institut für Rundfunktechnik of the RFA and the firm Philips of Eindhoven presented a process which was more elaborate than the simple superimposition of two images at projection; it consisted of scanning a screen of 1250 lines where each image occupies 625 lines of the screen, the even lines for one colour, the odd for the other. The firms JVC Sharp (Crystaltron), Fisher and Tektronics have also designed somewhat different processes.

Other firms and researchers have also tried to attain a more delicate stage, which is the realization of the illusion of three dimensions in colour; in fact, the use of complementary colours only produced grey and rendered the reconstitution of colour impossible. The reconstitution in question is possible using colourless **polarizing filters**, of which the polarization planes, being perpendicular, separate the images to the left and right without altering the colours. Polarizing spectacles with planes corresponding to those which were used at the separation of the images then have the same effect as the glasses described above: they only allow each eye to see one image.

All the same, this type of reconstitution is complex, for it requires on the one hand sets equipped with two **cathode tubes**, one for each image, two screens placed at 90°, and a semi-transparent mirror system which allows the images to be seen in superimposition; or alternatively, a screen with microscopic polarizing filters. In any case, this latter system which is evidently more simple, also requires 'active' polarizing glasses, connected to the viewer either by a wire or by an infrared or ultrasonic connection. In 1988 the price of the glasses alone (£1000 for those made by the American firm Stereographics Inc., for example) considerably limited the commercial expansion of the process.

Among the firms and researchers which featured during the 1980s as the leaders of the work on three-dimensional television, are the American firms Stereographics and Atari, the Frenchmen Christophe Odet from the INSA in Lyon (Ceravision process), Marc Chauvière and Jacques Guichard of the CNET much of the research is to do with special materials for the polarizing filters, such as a **zirconium titanate transparent ceramic** doped with lead and lanthanum (Stereographics), or else separating liquid crystal screens (JVC, Atari, etc.), or **separating mesh screens** (Tappan Printing, Japan).

Videodisc
Philips, Thomson-CSF, RCA, JVC, c.1972

A videodisc or disc video is a flexible or rigid plate of plastic material on which images are recorded and can be played back on a television screen using the appropriate reading device. This type of medium was patented by several firms at once starting with the Dutch Philips around 1972, at the same time as their reading system. Numerous other firms and private inventors during the 1950s studied and put forward prototypes of this sort; indeed, the videodisc is what might be called a development–invention, since it represents a spin-off of **magnetic recording onto tape**, this kind of medium being replaced by the disc. In the Philips process the principle is the same as that of television: a flux of electrons in proportion to the intensity of the message is emitted by the tube and recreates the image line by line. The disc is in fact cut with 45 000 grooves in a spiral, where the hollows are all of the same width and depth, varying only in length and spacing which constitute the information transcribed by a **laser pick-up head**. The laser messages are transmitted to a computer which converts them into electron fluxes.

In the Philips process the hard disc must be turned over for the other side to be read. However, in the process of the French firm Thomson-CSF, the disc is transparent and all that needs to be done to read one or the other side without handling is to change the focus of the laser beam. Moreover, with the disc being flexible, an equal distance is always kept between its surface and the pick-up head, which means there are no interruptions in the reading due to vibrations.

Finally, in the process of the American firm RCA, the reading takes place not using a laser but with a **polarized electrode** which transforms the variations in distance between itself and the disc into electric signals, depending on whether there is a hollow or not and on how long it is. It is appropriate to mention also the more recent method of the Japanese firm JVC, where the pick-up head is guided not by the groove itself but by an **optical tracking signal** incorporated into the video signal. These latter two processes are called **capacitive reading**.

Having prematurely aroused enthusiasm at its birth and inspired some often considerable financial investments, the videodisc suffered equally hasty unpopularity and only began to become widely used in the late 1980s, particularly in the professional domain (**data banks**). One part of its temporary unpopularity was to do with the reliability of the reading systems; another was linked to the size of the investment necessary for it to be widely used and to the unknown factor of its commercial profitability. However it was estimated at the end of the 1980s, 20 years after the invention of this means of recording data, that it had not yet reached its full expansion.

Videophone
AT&T, 1927; Reichspost, 1936; Chen, 1977

It is extremely difficult to assign either a date or a particular inventor to the videophone or **videophony**, that is, the linking of **telephone** to **television**. Indeed, there are numerous references to be found to the logical continuation of telephony in scientific literature, some even going back to the 19th century scientific illustrator Alfred Robida. All the same, a system remained to be designed which was capable of simul-

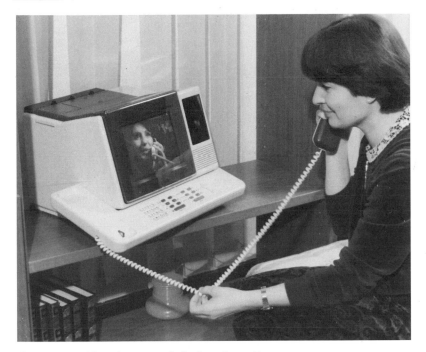

The prototype of the voice-operated telephone, *invented in 1985 and which will be operational during the 1990s. Since it has a totally smooth surface it is protected from vandalism.*

taneously transmitting and receiving images in parallel to sounds, something which was not going to be without essentially technological difficulty.

The oldest reference to a working videophone seems to be the conversation between the American Secretary of State for Commerce, Herbert Hoover, and Walter Sherman Gifford, president of the American Telephone and Telegraph Company (or AT&T). It took place in 1927 and enabled each speaker to see the other, with the help of what was in fact no more than the combination of television and telephone, with no specific equipment. The experiment was short-lived, at least in the United States.

However, in Germany, Reichspost set up a videophone service in 1936 with overhead cables between Berlin-Witzleben, Leipzig, Nuremberg and Hamburg. Communication took place not from household to household but from a special booth to a

similar booth. Transmission took place using a specially designed device, inspired by **Nipkow's model scanning system** (see p. 61). The image to be transmitted was imaged by 90 objective lenses in 1/25th of a second and consisted of 180 lines at reception (which shows that it was of poor quality compared to our television pictures nowadays).

The research in this field was not taken up again until the 1960s, when it was still based on the classic television–telephone connection. This happened particularly in Italy, Japan, and the USSR where there was a service which made use of the normal televisual network outside transmitting hours. In the United States the subscribers to the AT&T had access to a system called **Picture-phone** in 1971 (in Chicago and Pittsburgh to start with) but it was not successful commercially. Only a hundred models were still in use at the end of 1972 because it was expensive to use. It had already become apparent that the videophone had a future only on the condition that its development and cost could compete with those of the ordinary telephone. So it was necessary to return to a system of **digital transmission**. This consisted in translating the images into numbers in order to reduce them to compressible data which would be sent by the short channels of the telephone lines (or of the cable) to a receiver equipped with a decoding system which would re-transform the numerical data into images. The **analogue system** though, consists of simply sending the images point by point, according to the intensity of each point.

From 1977 the American W. Chen, firstly with Ford Aerospace, then with Compression Labs Inc., developed his research on the basis of the following principle: the image is broken down into **pixels** (for Picture Element) as for television, each pixel being defined by numerical values of **luminosity** and **colour**; out of 625 lines of 720 pixels, the 1987 technological consensus consisted in retaining only 576 'useful' ones, which meant that the picture was defined by 3 million 'bits' of luminosity and as many bits of colour. At the rate of 25 images per second it would be necessary to transmit some 166 million bits or 166 megabits per second, which exceeds the transmission capacity of **optical fibre**, if an image of quality comparable to television is desired. The tendency in 1988 was to sacrifice the quality of the image of the

feasibility of videophony by reducing the rate to 34 megabits. This is then a quantity compatible with the capacity of the telegraph network, which is 64 megabits per second, but only, of course, after the messages have been compressed.

This problem of the **compression of images** touched on the basis of a complex mathematical system which can be merely summarized here and which consists not of transmitting all of the data on each pixel, as if one was just re-applying analogue transmission in numerical transmission, but of transmitting only the difference between one pixel and the next. This problem has been the subject of research in France, at the European R.A.C.E. (Research in Advanced Communication Technology), the Swedish Televerket and numerous American and Japanese laboratories as well as at British Telecom. The ISDN (Integrated Services Digital Network) which British Telecom plans to have operational in certain parts of Edinburgh, UK, by 1992 will enable voice and picture information to be sent down public telephone lines at high speed. On 14 January 1991 the first UK public demonstration of a videophone took place, with a connection made between the Sheraton Hotel and the Editor's office at the *Scotsman* newspaper.

Also in Edinburgh, at the University, the development of a single chip video camera by Professor Peter Denyer and Dr David Renshaw has the potential to reduce significantly the cost of this technology.

Video recorder

Ampex, 1956, 1958; BBC, 1950; Sony, 1965; JVC 1976

The video recorder is a device to record the picture and sound signals from a television camera on **magnetic tape**. From 1956 to the 1980s, tape 2in wide ran at 15in per second past a rotating drum with **four heads**; sound and control signals were recorded on the two edges of the tape. Modern recorders use narrower tape with **helical scanning**. The first recorders were

developed for use by television stations; the first colour recorder was produced by Ampex in 1958)the VR1000B). The first domestic video recorder was produced by Sony in 1965, but the great influx of video recorders into homes could not begin until the easily-loaded **VCR (video cassette recorder)** was produced by Philips following research by teams in Eindhoven,

Netherlands and Vienna, Austria. The type of domestic video recorder most used now is the **VHS (Video Home System)** **format** launched by JVC in 1975. It accounts for 80% of sales in the world market.

Voice-operated telephone
CNET, 1988

Development of **voice-operated robots** at a time when the persistent vandalism to telephone booths was causing concern led the national French Centre for Studies in Telecommunications at Lannion in Brittany in 1988 to make a prototype of a telephone booth which had no visible telephone equipment. This prototype consists of a slot for a card to be inserted, a button for the dialling tone and a microphone. Out of the user's reach there is a microphone connected to a computer which is equipped with a memory of about 60 words, of which the key words are for emergencies, 'Police', 'Fire brigade', 'Send', 'Correction', in addition to 10 figures. All that is required for a normal call is to announce the figures one by one.

Wide screen cinema
Grimoin-Sanson, 1900; Chrétien, 1925

The origins of the projection of cinema films onto the wide screen, later defined as Cinemascope, are relatively old. Indeed, it was in 1900 that the Frenchman, Raoul Grimoin-Sanson had the idea of making a bank of ten cinema projectors to project a hand-coloured film onto a **full circle screen** with a 100-m circumference; the public were situated in the middle of the circle and could turn their heads right round without ever seeing the solution of continuity of the images. The films thus projected had of course been filmed in a broken-down panorama which was effected by the juxtaposition of sections, that is, with the cameras filming shots to be fixed together in sequence. This was one of the 'star attractions' at the Universal Exhibition in Paris. The idea was to be taken up again by the American Fred Waller under the name of Cinerama and on the decreased basis of three projectors and a curved screen (in fact, Waller was working on this from 1936). However this was rather a makeshift set-up which required bulky equipment and included a fundamental fault – that the three projected images overlapped at their adjoining edges. This problem would be solved by finding a way of wide-angle filming onto an adequately sized single film which was achieved by the French optician and astronomer, Henri Jacques Chrétien, when he invented the **anamorphic lens** in 1925. This compressed the image laterally at the filming stage and therefore, after correction during projection, allowed both a field much wider than that of the traditional image and the ability to project a single image onto the big screen. The invention was called the **Hypergonar** and it was the company Twentieth Century Fox who launched it in 1953 under the name Cinemascope. The image on Cinemascope was made on a film of four magnetic tracks.

Xerography

Chester Carbon, 1938; Xerox

Xerography is the most widely used method of photocopying. In 1938 the US physicist Chester Carbon devised the procedure whereby a photoconductive surface is charged and exposed to the image to be copied; the charge is lost except in the image areas. The secondary image is developed with a charged pigment which is transferred to copy paper and fixed by heat. He invented the name xerography (from the Greek words *xeros* meaning dry and *graphur* meaning to write) and patented the invention. He could not find a company to produce his invention until the Battelle Memorial Institute agreed to develop his idea in 1944, working with a small photographic firm called Haloid (which later changed its name to Xerox). The first photocopier, the Xerox 914, was brought into the market in 1950.

Electronics and Mathematics

Mathematics developed as a means of exploring the Universe using imaginary numbers. It allowed guesses to be made as to the distance between the Earth and the stars (which no one in ancient Greece imagined would ever be reached by man). With the passage of time, geometry became more refined so that it broke its ties with its father, Euclid; it began to describe a unique world which was almost inaccessible to the common imagination since it referred to such things as, for example, negative time.

When electronics first began it was still part of electricity, hardly anything more than an area of physics concentrating on one single particle, the electron. Towards the middle of the 20th century, however, electronics gave its services to mathematics in order to make the first electronic computers; first the top-secret Colossus I built in the UK during World War II to break the German Enigma codes, then, the famous monster ENIAC, which covered tens of square metres and weighed some 30 tons. From these beginnings followed the rest of computer development, given added impetus by the invention of the microchip.

The electronics–mathematics union was going to see one more great moment of glory, which was the eruption of the quantum theory in the explanation of otherwise incomprehensible electronic phenomena, such as the soliton state in molecular transistors or the attempted application of superconductivity to transistors. During the 1980s, scientists were seeking organic molecules which could perform the functions of electronic circuit elements, drifting towards philosophical speculation, such as the catastrophe theory and cosmological physics, which in a sense it absorbed. Indeed, after having been an instrument of physics, 'maths' ultimately became its container.

Computer

Abraham and Bloch, 1918; Eccles and Jordan, 1919; Bush, 1927;
Aiken, 1937; Eckert and Mauchly, 1946

The generic term 'computer' has been used since 1950. It applies to **mathematical instruments** which are to a certain extent self-governing due to their electronic circuits. These instruments which are also called **calculating machines** appeared several centuries ago; they could carry out distinct functions using gearwheels and exclusively mechanical devices.

Since the beginning of the 20th century numerous mathematicians had envisaged the use of electricity in calculating machines; it seems that it was the Frenchmen J. Abraham and E. Bloch who were the first in 1918 to have applied electricity to the working of a calculating machine, following the principle of **binary calculation**, where the numbers are expressed in series of two figures, 0 and 1, which correspond to the opening and closing of a circuit. Though sometimes attributed to the German Zuse, this idea was nevertheless brought to its first stage of maturity, which marks the birth of the **first generation** of electronic calculating machines, by the Englishmen W. H. Eccles and F. W. Jordan who in 1919 built the first electronic system. This was based on the **coupling of two electronic tubes**; the signal or information passed – or did not pass – in one direction, which corresponded to the entrance, and it went out – or not – in the other direction, which was that of the exit. This was the prototype of the bascule principle, also called **flip-flop**, which was later put into widespread use in the construction of digital computers.

By 1927 the American Vannevar Bush and his colleagues from the Massachusetts Institute of Technology built on this developing principle to make a **differential analyser**, which was undoubtedly the first reliable calculating machine; it could resolve a huge range of differential equations. It was carried to its maximum point of perfection within the bounds of the first generation of mathematical machines by the American Howard Aiken who built the IBM Mark 7 in 1937. The Mark 7 was extremely heavy, very cumbersome (it measured 15 m in length and was 2.4 m high), extremely complex (it contained 800 km of cables) but nevertheless very fast (it could do additions in 0.3 s and divisions in 12 s). It was very much an electronic machine but its actual relays were electromechanical.

The model which succeeded it in 1942, the ENIAC (Electronic Numerical Integrator and Computer), which was built by the Americans John Mauchly and John Eckert (with the collaboration of A. W. Burks, C. C. Chu, J. Davis, K. Sharpless and R. Shaw) was not much lighter (30 tonnes), but it was a thousand times faster, had a memory of 20 numbers and 10 letters and, finally, it was entirely electronic.

Finally completed in 1946 after numerous modifications since it was first made, the ENIAC, which is the ancestor of modern computers and the last of those of the first generation, comprised some primary and other secondary drawbacks. The secondary ones lay in the fact that it tended to overheat due to its 18 000 vacuum tubes (which required a complex air circulation inside the casing) and often broke down; it was difficult to repair and its energy consumption was very high (175kW). Its main drawbacks resulted from its small memory and, consequently, the difficulty in using it.

The next stage had been heralded by Alan Turing's work (see p. 84), and its aim was to make a machine with a big enough memory not only for storing a larger amount of data than the ENIAC, but also for recording the data during the operations, with a memory access time less than that of the operations themselves which was an essential condition for it to function quickly. Now, this was impossible with the technology used in the ENIAC: each memory relay, which consisted of a system of **magnetic cores** on stems, was controlled by a valve which was itself linked by innumerable connections to other valves;

increasing the memory ended up worsening the bulk and operational fragility of the whole thing. As for increasing the operating speed, this was impossible with the equipment available.

At this time two inventions, **core stores** and the **transistor** (see p. 84) made an important impact. The former, which were followed by even better improved memories, increased the ease of use considerably, whilst the latter reduced the bulk. The **second generation** of computers was inaugurated by Remington Rand's UNIVAC, the EDVAC of the University of Pennsylvania and of the U.S. Army, the SEAC, in the National Bureau of Standards of Washington, and the EDSAC in the mathematics laboratory in Cambridge, which were all faster, they performed better, were less cumbersome and more reliable.

The **third generation** of computers began during the 1960s with the invention of **integrated circuits**; it consisted of electro-

nic microcircuits based on transistors and resistors. These microcircuits, which were soldered to rigid supports, enabled considerably less cable to be used; thus **miniaturization** in computers was introduced. The capacity of these increased a great deal although the dimensions of the casings were several hundred times smaller. The first integrated circuit was made in 1959 by Texas Instruments and the first one was commercialized in 1965 by Fairchild Semiconductors. The advent of the **disc memories** was to increase the capacity of computers further, at the same time as mass production progressively reduced their cost.

The generic term computer refers to three types of apparatus. **Analogue computers** represent values or numbers by physical quantities; they resolve problems on the basis of continuous variables and not discontinuous units. **Digital computers** calculate specifically or discretely using neither variables nor degrees, but responding only to exact signals of the flip-flop type; their data is represented by letters, figures and symbols while being confined to a finite number of alternatives. Finally, **hybrid computers** combine the above two types and function in a similar way to the human brain.

The first nuclear physics problem submitted to the ENIAC would have taken a century to resolve by hand, or two weeks with a mechanical machine of the Babbage type; the ENIAC completed it in two hours.

Computer 'vaccines'

IDC Ltd., Sectra, Plus Development, 1988

After several incidents akin to sabotage were reported in the computer systems of large companies between 1986 and 1988, the affected firms began to search for a **protection system for recorded data**; these systems are called 'vaccines', the functional defects being called 'viruses'. Indeed, being introduced surreptitiously into the computer programs, they are liable to damage if not actually destroy them, and they are 'infectious', since they are also liable to take over other systems during a memory transfer.

These 'vaccines' actually sit on **memory**

input access codes on computer discs, whether these are hard discs or not; in one of these 'vaccines', which is called 'Canary' and is produced by the British firm IDC.,

---

The scares which have led to the making of security systems have often been justified. It was in this way that someone realized that a young German had had access to about 30 computers belonging to the US military services in Germany and that he had thus had the chance of destroying, breaking and also violating them.

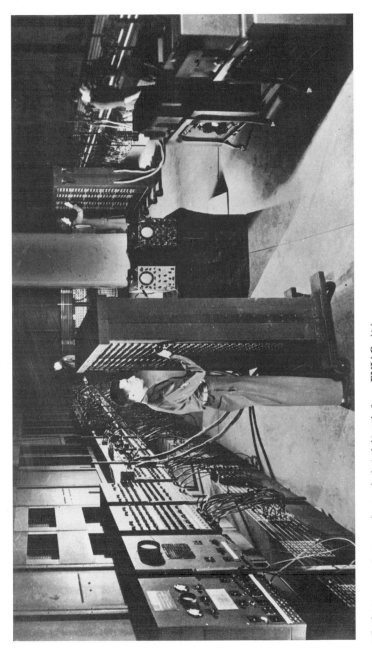

The first-ever computer, an almost archaeological object, the famous **ENIAC** which was built in 1946 at the University of Pennsylvania. It occupied all the cupboards in the room shown here and gave out heat which meant the very powerful ventilation systems were needed.

any interference by an operator not in possession of the access code automatically replaces the picture of a live canary on the unspoilt disc with one of a dead canary, which constitutes a warning system. The 'T-cell' system of the Swedish firm Linköping consists of such a complex **logic code** that its integrity can only be altered by very powerful computers, and so by a recognized and highly organized determination to sabotage. The 'Passport Plus' system of the American company Plus Development in Milpitas, California, consists of a mobile hard disc which is taken out after use. Other computer firms have also made similar systems.

Electronic typewriter
Olivetti, Casio, 1978

The first electronic typewriter, which was equipped with a **memory** capable of storing text, was produced in 1978 by two firms simultaneously, the Italian Olivetti and the Japanese Casio.

Geometry
Riemann, 1854; Beltrami, 1868; Klein, 1872; Lie, 1886; Cartan, 1894; Henri Poincaré, 1873–1912

The major event to appear in geometry at the end of the 19th century was **non-Euclidian geometry**, which was invented almost by chance in 1773 by the Italian Girolamo Saccheri, and then structured between 1836 and 1840 by the Russian Nicolas Lobachevski and the Hungarian János Bolyai. One of the characteristics peculiar to this type of research at the time was that since scientists published little, they had relatively little contact with each other. It was in this way that Lobachevski and Bolyai arrived independently at the same concepts which instituted the notion of the curvature of space, which the German Bernhard Riemann also calculated in 1854.

Of these three great geometricians and mathematicians it would seem to be Riemann who drew the most advanced conclusions. Riemann worked assiduously at the problem of the **nature of space** and he invented the concept which was later taken up by Albert Einstein: space is not a neutral environment but one which in-teracts with the objects which are immersed in it. He maintained that it was impossible to draw a line parallel to a straight line. For a time his posthumously published writings remained unrecognized. During the following years the Italian Eugenio Beltrami demonstrated that if the distance and the angle as measured in a Euclidian plan are simply thought of as the functions which have provided the distance and the angle as measured in spherical space, then one returns to the Lobachevski system. He thus made a transition between Euclidian and non-Euclidian geometry, which was made even neater by the German Felix Klein in 1872. He in fact invented the principle according to which there are **invariant** properties of space corresponding to every group of transformations found in space. Several types of geometry are distinguished in this way, each one of which is characterized by a group of transformations and invariants: the **projective** which conserves the notion of point and straight line; the **affine**, which conserves the straight line but not the lengths and sizes of angles; the **metric** which conserves distances; the **Euclidian**, which conserves length, dimension, size of angles and forms figures; the **non-Euclidians** which are shared between two groups: the **hyperbolic metric**, which conserves a real conic as an invariant, and the **parabolic metric**, which conserves the size of angles.

Two great mathematicians, Lie and Cartan, followed Klein by opening up the new domain of non-Euclidian geometry. Sophus Lie, a Norwegian who worked at Leipzig applied himself to defining the contact transformations between spheres and straight lines in space. His analytical formula founded what is now called **Lie algebra**. The Frenchman Elie Cartan refined the exploratory instruments in this domain yet further which were becoming more and more closely linked to algebra; in 1894 he gave a classification of all Lie's groups, work which he was to complete in 1914 with the development of Lie algebra. Cartan is also famous for having devised general geometries which comprised Riemannian geometry.

A special mention should be made of the dominant mathematician of his age, Henri Poincaré, who applied himself to the application of differential equations in **mathematical physics** and in **celestial mechanics**. As a 'specialist' of the **functions** of which he had discovered two classes, he was the first to study the analytical functions which had several complex variables. This would lead between 1881 and 1911 to **analytical geometry**, which consisted of the theory of varieties and of analytical spaces. His work comprised nearly 500 scientific papers; only a glimpse of his conclusions on the practice of mathematical theories can be outlined here. Poincaré seems to be the first scientist to have realized and declared that neither geometry nor mathematics could constitute a representation of the world; they were simply a tool for working towards knowledge of the world. In fact, Poincaré differed from his predecessors in that he often studied the application of geometry and mathematics during analyses of physical problems for example the electrodynamics of moving bodies, the vibrations of membranes, the cooling of a solid and planetary orbits. Poincaré was the first geometrician and mathematician to have integrated his knowledge into the field of **theoretical physics**.

This history of geometrical inventions from 1850 is necessarily extremely brief, and the names and dates quoted are just the landmarks. The history of geometry is one of an unbroken succession of inventions and only a vast work could do justice to its many heroes.

Geometry's contribution to technology is inestimable. From mechanics to astronomy, from architecture to theoretical physics, its spin-offs are incalculable.

> Surprise seized the scientific world when Albert Einstein published his Theory of Special Relativity in 1905. Poincaré had been expected to formulate this key theory, for he had mastered the conceptual aspect of it as well as the instruments of analysis, and he was an incomparably powerful mathematician. Perhaps the audacity of the theory of relativity made him hesitate.

Klystron

Varian brothers, 1937

The klystron is one of the major inventions of the 20th century; it is used in radar, long distance telephone communications, broadcast television and cable television, satellite communications, particle accelerators and radiotherapy apparatus. It consists largely of an electronic wave **amplifier** and **accelerator**. When it was invented in 1937 by the brothers Russell Harrison Varian and Sigurd Fergus Varian, Americans of

Irish origin, the klystron was in the form of a **cathode ray tube** inside which an electric current heats a filament which in turn heats the cathode, the surface of which is treated in such a way that at a given temperature it emits electrons. These are attracted by the positive anode which they reach after passing through a special cavity in the tube, called a resonant cavity; here high-frequency microwaves interact with the electrons which, still in this resonant cavity, tend to group together with some accelerating and some slowing down. The electrons pass into second resonant cavity; they get more excited and high-frequency waves are generated in it. The generated high-frequency microwaves are then emitted at the other end of the tube.

Since neither of the Varian brothers had any fundamental knowledge of electronics, the Department of Physics at Stanford University, which had been given the task of developing the invention by the Sperry Gyroscope Co., which had purchased the patent, asked for the help of some technicians. These men, Edward Ginston and John Woodyard, made the first generator for short waves which were required for radar (see p. 153) using the klystron. The first klystron which was derived from the **cavity resonator** cost $100 to make, paid for by the president of Stanford University.

Microchip
Saint Clair Kilby, 1958

In electronics, the microchip is an instrument which contains several **transistors** (see p. 84). It was invented on 12 September 1958 by the American Jack Saint Clair Kilby who worked at Texas Instruments. This prototype only contained a few transistors and capacitors, which were on one surface, and the size of the whole was much smaller than the same components if they had been assembled separately. The transistor, which was already involved in the race to miniaturization, did not resist competition for long: after being adopted first by the US Army for the electronic circuits in rockets, the microchip passed into the public domain in 1973. Saint Clair Kilby designed a microchip calculator which was as powerful as the electromechanical machines of the time but which could be held in the hand. This was the 'Data-Math' which gave rise to quite a revolution, firstly because of its relatively low price ($149.50) and also because it was easy to handle and very efficient (four operations in 5s). The microchip itself has been made smaller since then and at the end of the 1980s some of them comprised up to a million transistors.

Microcomputer
Ahl, 1974; Wozniak, 1975

Two essential factors worked towards the invention of the microcomputer: the development of the **microprocessor** and the annoyance cause to computer users by the fact that access to the big computers was so difficult.

The first prototype was made by David Ahl, an employee of DEC, which is the second computer firm in the world after IBM. It was the size of a large television set (because among other things it included a television screen) and a keyboard. When presented to DEC., it hardly aroused any interest at all.

In 1975 Steve Wozniak designed another prototype independently and had no more success. The Apple II was launched in 1977, followed by the machine of great importance to non-specialists, the Apple Mac, in 1984.

Molecular transistors

Lyons, McDiarmid, Mott, 1978; Friend and Cambridge University, 1988

A molecular transistor is the name for a transistor, the active elements of which are made out of **carbon-based materials** and are therefore organic. This idea was evoked several times in the scientific literature at the end of the 1980s. The pioneers were researchers around 1978, such as the American L. E. Lyons and the British Hugh McDiarmid and Neville Mott. The first experiments took place using transistors with diodes of organic materials doped with oxygen; they were interesting in theory but disappointing in practice. The first convincing results came from the Englishman Richard Friend and a team from Cambridge University. Making the most of new chemical processes, in 1988 Friend made some diodes out of **polyacetylene**, an organic polymer. In this way he obtained results which he claims are a thousand times more convincing than those of preceding prototypes consisting of organic elements—for they show an amazing output of energy, approximately 5% efficiency.

In classical semi-conductors, based on silicons and gallium arsenide, the electrons which circulate in the materials do so in two bands: a **valence band** where the electrons are linked to atoms and a **conduction band** where they are free to move around and convey the electricity. There is a difference in energy between the two bands called the 'band gap'. When the current circulates some electrons from the valence band are torn away and pass into the conduction band, thus leaving spaces in the valence band. As the polyacetylene is based on a chain of alternately single- and double-linked carbon atoms, it turns out that the valence and conduction bands are superim-posed and the gap between the two represents an energy difference of only 1.5 electron volts. All the same, this gap is not as distinct as in the classical transistors. The redistribution of the charges induces a state which is mid-way between the valence and conduction bands. So in theory the organic transistors should be quicker and more economical than the classical transis-

Molecular transistors should not be confused with the organic transistors which were devised by certain researchers at the end of the 1980s and which were to consist of electrochemical circuits set up between **living cells**, such as bacterial or tissue cultures. This principle, which is highly speculative, is based on the fact that a living cell is capable of storing a considerably larger number of pieces of information than any inanimate substance, at least in the current state of scientific knowledge. An electronic system using living cells would function on the basis of impulses being given when there was an exchange of specific chemical substances or of exactly measured electrical frequencies. Ultimately it would even be capable of doing the same operations as the brain, but it would also be capable of repeating them indefinitely. A system such as this would constitute the closest ever approximation to the living brain. It could only function however in very narrowly defined circumstances (temperature, pH (acidity) etc.) and possibly for a limited time. It implies knowledge of electrochemical exchanges which would be even more delicate than those actually present in living organisms.

tors. Moreover, the organic transistors display a specific peculiarity: the electrons are pushed from the valence band to the conduction band by **photons**, providing that these are of an energy higher than that of the band gap. In other words, when subjected to **infrared** radiation the polymer diodes amplify the current, a property which, furthermore, had been noted in 1978.

The practical perspectives of this invention are so diverse that at the end of the 1980s it was predicted that a general replacement of classical transistors by molecular transistors would take place.

Monocrystalline superconductors

Fujitsu, 1988

Any current of electrons in a conducting material experiences a loss of energy due to the impact of electrons against the parts of the crystal lattices over which they cross. The slowing down of the flux of electrons is greater or smaller depending on the configuration of the lattice, but under normal conditions it involves a loss proportional to an emission of heat, which is explained by the energy transformation principle. During the 1980s a renewal of interest in the **superconductivity** phenomenon led to the discovery of materials which at very low temperatures offer no resistance to the passage of electrons. The phenomenon, which can be explained in a very simplified way by the alteration of the lattice configuration, had been discovered in 1911 by the Dutchman Kamerlingh Onnes and then thought of for half a century as a laboratory curiosity of no practical interest. In fact, attaining temperatures bordering **absolute zero**, $-273°C$, costs much more in electricity than is profitable. Then the discovery in 1987 of materials which were superconductors at temperatures well above absolute zero stimulated research and led to the discovery of the superconductive properties of certain **copper oxides** in particular, the properties of which were all the more interesting for the fact that they were manifest at temperatures around $0°C$. This important discovery was to be worth a Nobel prize almost at once to the Swiss researchers Bednorz and Müller.

One of the problems met with in the study of superconductivity is that of the thickness of the superconductive materials. In fact, materials which comprise several thicknesses of lattices tend to dissipate the energy more easily than materials which consist of one single crystalline layer. Starting from this fact, the Fujitsu laboratories, from the Japanese electronics firm of the same name, designed a technique of making crystal superconductors of a thickness of one single layer of crystal, called monocrystalline for this reason, and measuring about 0.33 μm. The material in question which can be used in the manufacture of electronic components is a composite of bismuth, strontium, calcium, copper and oxygen. It belongs in the group of **ceramics**. In electronics, such a material enables the construction of very large integrated circuits with minimal energy loss and unequalled rapidity. It also allows manufacture of ultrasensitive devices, eg the SQUID (Superconducting Quantum Interference Device), which can measure the very fine variations in a magnetic field.

Mono-electronic transistors

AT&T, 1988

Some transistors, so sensitive that the passage of a single electron is sufficient to induce an electric current, were invented by the engineers of the American AT&T laboratories in 1988. As they can function at very low temperatures they can be used as **electrometers** for measuring charges to an accuracy of 1% of the charge of an electron (charge on electron= 1.6×10^{-19} coulombs).

Pen Writer (Manual graphics computer)

Scriptel, 1988

The first computerized electronic system to enable handwritten graphics to be transferred and stored in memory was commercialized in 1988 by its inventor, the American firm Scriptel. This device, which can be called a manual graphics computer and which is called 'Pen Writer' by its producers, consists of a screen on which one can write a message by hand which is then stored in memory. One version of this system can also transcribe messages consisting of words and figures into standard characters. This system, which has to be linked up to an IBM-compatible computer, has a choice of three screens: one fluorescent liquid crystal, one electroluminescent and the third, neon gas plasma. It is possible to erase certain parts of the graphic messages.

The Pen Writer or pen for writing on a screen, completed in 1988. Like the voice-operated computer, it should considerably change relations between the public and computers; for example, it enables the signing of a letter from a distance.

Photoelectric cell
Elster and Geitel, 1896

One of the devices used the most in modern technology, from astronomical photometry to television, in methods of surveillance, the location of colour in printing, and solar cells, was invented in 1896 by the Germans Julius Elster and Hans F. Geitel. Four years later it progressed from laboratory apparatus to practical application. A photoelectric cell consists of an **electronic tube** of which the cathode is sensitive to light rays; when it is hit by photons it is stimulated and thus emits electrons which are captured by the anode. The current thus produced is proportional to the intensity of the light.

The photoelectric effect itself was described in 1837 but it was 'rediscovered' and still only partially understood when the German Rudolf Hertz noticed in 1887 that ultraviolet light can alter the lowest voltage required to make a spark between two electrodes. Subsequently it was deduced that the cathode was emitting 'discrete' negative rays, called **cathode rays**. The identity of these rays with the electrons was established in 1900 by the German Philipp Anton Edward Lenard. In the meantime Elster and Geiter had thought of designing a light detector based on this principle. In 1902 the phenomenon in use began to be a little better understood; it was also discovered that the maximum velocity of the emitted electrons was independent of the intensity of light received by the cathode. But it was only in 1905 that Albert Einstein explained it completely by postulating that light can be considered to consist both of waves and **quanta** (or packets) of energy, which induce the emission of the cathodic electrons.

Set theory
Cantor and Dedekind, 1876–79; Schröder, 1877; Peano, 1895; Zermelo and Fraenkel, 1904–10, Borel, 1916–25

The set theory is a branch of **axiomatic mathematics** which allows the treatment of **discontinuity** and which can be presented as symmetrical to the other branch, **topology**, which allows the treatment of **continuity**. Within this perspective the number, being discontinuous, is the object of the set theory, while space, being continuous is that of topology. It is a creation, and therefore an invention, which is used in numerous branches of mathematics and particularly in **probability calculation** and in **computer science**.

In order to understand how it came about, it must be seen in two stages: the first, which can be called the intuitive, and the second which came later and was specifically axiomatic. The intuitive stage, that is, that of the actual invention is due to George Cantor (a German of Russian origin), and his fellow countryman Richard Dedekind. Following discussions with Dedekind, it was Cantor who, between 1876 and 1879 formulated the set theory: 'I call a set a whole collection M of objects m which are distinct from our thought or our perception.' It will be noted in passing that the formulation in German is rather ambiguous, and it must be clarified that the objects m are actually distinct *in* our thought and perception. These objects were to be subsequently defined under the name of elements of the set. Therefore the term M has been replaced by E.

A set E is defined by being written, and it is appropriate to cite the invention in 1877 of the inclusion sign $\subset$ by the German mathematician and logician Ernst Schöder, which differs from the membership sign $\in$, which is quite separate and was invented in 1895 by the Italian Guiseppe Peano. If $A \subset E$ is written, it means that E includes A, but if one writes $x \in E$, that signifies that the element x belongs to E. Subsequently the sign ⊓ was invented, which indicates that one sign is distinct from another.

The formalism of this writing had a novel advantage: the possibility of describing discontinuous quantities and of finding the **laws of composition**. These were to lead up to the formulation of an '**algebra of structures** with morphisms, groups, ring monoids, bodies, isomorphism, modules, vectorial spaces, etc.' (A. Aaron-Upinsky).

In 1904 the axiom outlined by Cantor took shape and allowed, for example, the German Ernst Zermelo to set out the **axiom of choice**, also called **Zermelo's axiom**, which shows that every set is well ordered. It was the German Adolf Abraham Fraenkel at the beginning of the century however who put the finishing touches to the axiom of the set theory, which since then has been referred to under the sign ZF, the initials of the two mathe-

maticians. This axiom did not just constitute a formalism; together with the **constructibility axiom** which is implied in Zermelo's axiom, it was to enable paradoxes to be avoided.

The set theory is actually ruled by three notions. The first is that a set is considered as a finite collection; the second, that it consists of a collection of arbitrary sub-sets, and the third, that the set is really the abstraction of the notion of property, since it consists of a collection of objects of a given property. Until Zermelo and Fraenkel an argument could satisfy one of the three notions above but not all three at once. The ZF axiom assigned rules to the definition of structures.

This axiom was to be improved between 1916 and 1925 by the Frenchman Léon Borel, a top class logician and mathematician who refined the set theory by applying to it the principles of the **theory of functions**.

The set theory had obvious potential to be applied in computer science (since the element can be paralleled to a piece of information and the set to the file).

Superheterodyne receiver
Armstrong, 1917

Heterodyne, in electronics, is the name given to a device which produces pure or modulated **high-frequency oscillations** by combining the local alternating current with electromagnetic waves; in **radio reception** it acts as an **amplifier** (see p. 54). The results from the first heterodyne receivers at the beginning of radiophony were variable, given that the equipment was designed for low frequencies. In 1917, the American Edwin Howard Armstrong

thought of changing the signals into high frequency which was easier to amplify. This variable frequency receiver enables the listener or viewer to search for stations by pushing a single button. Westinghouse acquired the patent for the superheterodyne receiver for $500 000, a considerable sum at the time. The principle is employed in most radio and television receivers today.

Transistors

Bardeen, Brattain, Shockley 1948

The invention of the transistor in 1948 started a new era in electronics technology. Whole new industries are founded on the application of the transistor: modern computers, space flight, electronic musical instruments. Transistors detect, rectify, switch, and amplify electric currents, functions which, before the invention of the transistor, were performed by vacuum tubes such as triodes and other large devices. Many millions of transistors can now be made on a finger nail-sized piece of semiconductor.

Three Americans—John Bardeen, Walter Brattain, and William Shockley— working at the Bell Telephone Laboratories of AT&T, demonstrated the transistor in 1948. The first transistor radio, the Sony TR-55, was produced in 1955. The three American inventors were awarded the Nobel prize for Physics in 1956.

Transistors depend on the properties of semiconductors for their action. Silicon is now the most common semiconductor material though germanium was the first one used. Now some other materials have been produced for special purposes. A semiconductor is a material with electrical conductor properties intermediate between metals (good conductors) and insulators (non-conductors). Normally small quantities of impurities are used to dope the pure silicon of germanium to produce desired properties making them either 'electron rich' n-type or with a affinity for electrons, p-type. Placing a piece of n-type material in contact with a piece of p-type produces a diode, a device which will allow electrons to flow preferentially in one direction. A sandwich of two n-type layers with a p-type in between, or two p-types with an n-type between, forms a junction transistor. A low voltage signal between the emitter and the base (the middle layer) causes a current to flow between the collector and the base through an external circuit. The low voltage signal is amplified because the output resistance of the collection base junction is several hundred times bigger than the emitter base resistance.

Many different structures are now used to produce transistors with a variety of properties optimal for particular uses. These have not only replaced valves or vacuum for nearly every purpose but perform new functions. When the firm Westinghouse invented the **thin film transistor** it was possible to further reduce the volume. A simple film of Mylar was sufficient to support the three elements of the transistor and it became trivial to reduce the transistor to practically two dimensions and an area of less than one square millimetre. The path towards miniaturization was started. The first products to benefit were radios, then hearing aids, computers, and cameras.

Turing machine

Turing, 1936

One of the key concepts behind the **theory of automatic calculations** and the **theory of computing** was invented in 1936 by the English mathematician Alan Turing – the **Turing machine**. It consists of a machine which would be capable of accomplishing any task at all; in theoretical terms it is a logical machine capable of calculating any calculable number. It would work by scanning a band containing a theoretically unlimited number of compartments, each one of which would contain a finite number of instructions in the form of symbols. The great feature of this machine is that it would also be capable of itself carrying the information in the compartments, according to the task in hand, which would be the alteration of the information contained in the compartments. The Turing machine,

which is still referred to in work on computing and logic, has subsequently been shown to be feasible by the creation of **feedback** mechanisms.

The Turing machine, which was never built and is to this day a hypothetical prototype, was invented by the scientist in response to the German mathematician Kurt Gödel's famous paradox, which said that no mathematical system could be both coherent and complete. In order to show the validity of his paradox he laid down formulae, the essence of which can be summarized as, 'This postulate can never be proved'. So Turing designed a machine which could invent data as required whilst dealing with each task. Indeed for him, the constant solution of mathematical problems required an infinite number of new ideas. Turing, who was one of the greatest mathematicians of all time, committed suicide in 1953 following a prosecution for an alleged homosexual offence.

Voice-operated computer
Martins and Cox, EMI-Threshold, 1973

The first computer which could understand speech and then answer in a **synthetic voice** was made in 1975 by the British firm EMI-Threshold following the invention by the Americans Thomas B. Martins and R. B. Cox which was made in 1973.

The understanding of speech is based on the fact that the voice consists of a series of **vibrations** which can be broken down by the mathematical process of **Fourier transformation** into several **sinusoidal components**; a system of filters allows the **frequency** and **energy** of each point of the sinusoid to then be measured and for these to be translated into **binary language**. Each point corresponds to a series of signals (1 and 0) which reflect the absence or presence of 32 grouped phonetic character-

Before the end of the 20th century the voice-operated computer ought to cause a spectacular modification in the relationship between society and computer science.

istics. These follow the **linguistic** principle of five families: vowels, consonants, long pauses and short pauses and plosives. The computer equipped with this kind of analysis has a memory of 220 words of whatever language is required. Each word in then split up into 16 equal intervals which each correspond to $16 \times 32 = 512$ bits. In order to learn the voice and the diction of its operator, the computer records the same word ten times in a row and extracts an average speech pattern, which enables it to identify the spoken word (obviously using a restricted vocabulary), even if the voice of the operator varies in tone or delivery. The average images of each word are recorded on cassette and can be re-used by the machine itself to make a reply.

Energy and Mechanics

The history of human inventiveness in the domain of energy since 1850 is marked by one major phenomenon: the development of the practical application of atomic energy. It is a major and spectacular phenomenon, for during the 1930s most scientists still had their doubts about the possibility of one day being able to exploit such energy. Gustave Le Bon's short pronouncement on this subject in 1920 was hardly based on serious technical data. In 1942 however, the Italian Enrico Fermi discovered nuclear reactions produced by slow neutrons in the first atomic reactor. Research continues on the 'eternal' energy of a fusion reactor. Fission reactors compared to this would be like a car in comparison with a jet aeroplane. It is worked on assiduously but is not yet a reality.

During the 1980s superconductivity arrived to change the landscape of world energy further: certain copper oxide composites enabled apparently sensible and considerable savings in electricity to be made by reducing the resistance of the conductive elements. An electric car battery would in theory last for months.

At the same time, public fears concerning some of the problems of nuclear energy have set various productive initiatives in motion, leading to the development of ways to exploit solar energy. These fears were manifested themselves first as a real uneasiness about atomic power stations; the inevitable deadline represented by the finite supply of fossil fuels remained. Several nuclear accidents, the most famous being those at Windscale in England, Three Mile Island in the United States and Chernobyl in the USSR, have given rise in the 1980s to a unanimous crisis of conscience over the dangers of nuclear energy. The 1973 oil crisis served to remind the industrial world of its excessive dependence on the producer countries. Research on solar reactors, which in the beginning was looked upon with disdain by specialists, began to bear more and more fruits and it is likely that in the 21st century solar energy will be used on a considerably larger scale than it is at the present time.

Aerodynamics

Tsiolkovski, 1892–6; Citroën, Chrysler, 1934

Problems caused by the **air resistance** of a moving object have been experienced since the birth of **ballistics** (see *Great Inventions Through History*) but the lessons learned were unheeded until long after the vehicles operating on ground, sea and in the air had achieved speeds which resulted in such problems.

The first theoretician on the subject was Konstantin Eduardovich Tsiolkovski, a pioneer who between 1892 and 1896 built a great number of **fan engines** intending to define mathematically the forces of friction which are exerted on the surface of a vehicle. Tsiolkovski in fact studied a rapid airship prototype. Neither aeroplanes, cars nor boats took aerodynamics into account when they were being designed, with a few near exceptions such as the Belgian Camille Jenatzy's vehicle 'The Never Happy' which beat the 100 km/h record in 1899 with a chassis in the form of a shell. The first motor vehicles to attempt to reduce the resistance of air to forward motion were simultaneously André Citroën's 7A, the first front-wheel drive, and the Chrysler Airflow. Thenceforth the whole of the automobile and aviation industry was won over by the concern for aerodynamics. An excess of enthusiasm then spread to objects destined never to move, such as teapots, or only to move at very low speeds, such as irons.

A study of aerodynamics in a motorbike. Sketch of a Honda prototype.

Atomic energy

Einstein, 1907; Hahn, Meitner, Strassman, Bohr, 1939;
Fermi, 1942

The notion of **fission of the atom** proceeds as much from intellectual analysis as from chance discovery; so it is not an invention in the full understanding of the word; nevertheless, it cannot comfortably be excluded from the domain of inventions. Its seed was contained in Albert Einstein's famous equation published in 1907 where he compared **energy** and **matter** ($E = mc^2$). The possibility of making use of atomic energy remained however only intellectual speculation. One day in 1938 the Germans Otto Hahn, Lise Meitner and Fritz Strassman discovered that **uranium**, when bombarded with slow or fast **neutrons**, would break down into two other elements, **barium** and **krypton**, accompanied by an energy release of a previously unsuspected magnitude (200 million electronvolts). This really was a 'chance' discovery, the significance of which only Lise Meitner realized (see p. 217).

On the other hand, actual atomic energy, whether peaceful or military, was the result of an invention in the traditional sense and, more accurately, the peaceful use ensued from the military use. Indeed, peaceful atomic energy could never have been exploited if it were not for the research which led to the manufacture of the first **atomic bomb**. This allowed three essential factors to be clarified. The first was the nature of the element which could be used as the initiator for the energy-releasing **chain reactions**. After Hahn's discovery, which was extended by Bohr, it turned out to be uranium, and more precisely uranium-235, which was present in small quantities with uranium-238. Then after the discovery of neptunium it was understood that **plutonium-239** could also be used. However it took several years before the technology of uranium was mastered. Even when the first atomic bomb had begun to be made it was not known if the more suitable nuclear fuel would be uranium or plutonium.

It was the Italian Enrico Fermi who built the first known **nuclear reactor**, at the University of Chicago in the United States; this reactor reached the critical point on 6 December 1942. It functioned using **graphite** (400 tonnes), **uranium metal** (6 tonnes), and **uranium oxide** (50 tonnes) with control rods made of **cadmium**. The large quantity of uranium oxide is explained by the fact that it was natural oxide or **yellow cake**. The graphite served as **moderator** (to slow down the reaction). This reactor was successful as far as it went although it had the serious defect of having no **cooling system** (which exposed the operators to great danger should the reactor have suddenly gone wrong), and it had to be taken apart (not without danger) in order to get at the plutonium encrusted in the graphite moderator. Plutonium is actually a product of uranium fission, and the most dangerous of all.

This reactor enabled progress to be made in the knowledge of the second factor, concerning the **form of uranium** which would be the most profitable (it is known today to be **uranium dioxide**). The third and most difficult factor was the **metallurgy of uranium** which requires heavy, dangerous and expensive installations. It was not actually until after 1945 that the United States, despite highly competent teams and individuals, succeeded in obtaining a sufficient quantity to use to make the first A-bomb.

Fermi can be thought of as the real inventor of the atomic reactor, and the peaceful use of nuclear energy. The control system which he created consisted of producing a critical mass of uranium; as soon as the reaction had been set in motion, the uranium rods would be pulled out in order

In 1948 France began to use the first research reactor which functioned on natural uranium, with 'heavy' water (water containing a heavy isotope of hydrogen) as moderator.

to avoid overheating or, in the worst case, a runaway reaction. It was a system which had the additional merit of proving that the critical point could be obtained more easily than had been feared (this was a notable encouragement also to the engineers involved in the manufacture of the bomb).

The first known reactor to produce electricity was that of the Argonne National Laboratory which was put into use in 1951. It was called a **thermal fast neutron** reactor, without a moderator, which was based on the principle of the use of rods composed of enriched uranium (U-238 + U-235) at the centre and with natural uranium (U-238) on the outside. It had a considerable advantage in that it enriched the external part itself, transforming it into fissile plutonium. This type functioned on two speeds of neutrons (either slowed down fast neutrons, or slow neutrons) in order

also to initiate the fission of the envelope of U-238. However, this envelope picks up só-called **resonance neutrons** which transform it into **neptunium**, then into fissile plutonium-239. Thus, the uranium rod is theoretically entirely 'consumable' which increases its profitability considerably.

The electricity was produced by the harnessing of the heat, first of all by a liquid metal exchanger (**sodium–potassium**), then by a classical steam turbine.

> When the reactor at the University of Chicago began to function, Fermi telephoned his colleague James B. Conant at Harvard University to give him the following secret message: 'Jim, I think that it will interest you to know that the first Italian navigator has landed in the United States.'

Diesel motor

Diesel, 1892 and 1893

Thirty years after Beau and Rochas had established the principle of the four-stroke cycle in the internal combustion engine (see *Great Inventions Through History*), the German Rudolf Diesel, stimulated by the poor thermal efficiency of this motor tried to improve on it. He tried to apply Sadi Carnot's principle, according to which the output from a reciprocal or reversible movement motor depends on the temperature at which it operates, and which is one of the two basic principles of **thermodynamics**. Published in 1824 in a brochure entitled *Reflections on the motor power of fire and the correct machines to develop this power*, this principle unfortunately went by unnoticed at the time, when only about 50 motors were made each year in France and when such considerations as output were too far in advance of what was expected of mechanical power. It was only after more than half a century when Carnot's brother republished his *Reflections* that homage was paid to the genius of the physicist.

One of Carnot's leading ideas which was going to influence Diesel the most was that

the **piston cycle** had to modified, for that was what caused the heat loss. In a first patent, which was submitted in 1892 and revised in 1893, Diesel proposed to obtain the necessary heat for the air–fuel mixture in the classical combustion engine to burn, not by using a spark but by a very high **compression** of air alone which would be enough to bring the air to the required temperature. The fuel then turns to vapour in the cylinder and the pressure of the hot gases pushes the piston. Up to this point the principle of the diesel engine is not much different from the four-stroke engine, except that the spark is removed. But in other respects Diesel modified the design of his engine in such a way that the compression of air took place outside the cylinder; thus it was the compressed air which injected directly into the cylinder. Moreover, the injection of compressed air pushes the gases from the burnt fuel through openings at the base of the cylinders; the track of the piston is reduced for it no longer has the space to descend to the base of the cylinder.

Thus Diesel replaced the four-stroke cycle with the **two-stroke cycle** and was drawing closer to the **incomplete cycle engine** proposed by Carnot where the heat loss was reduced by the further reduction of the piston track.

Diesel also introduced a significant factor of economy: his engine could function on **unrefined fuels**. In 1896 the Krupp conglomerate became interested in it and attempted to make a coal Diesel engine which was abandoned. But the following year an engine functioning on unrefined oil gave satisfactory results.

The Diesel motor rapidly proved its **reliability** and the superiority of its **output**. From the point of view of reliability, it is easy to manufacture, strong and hardly ever breaks down; from the point of view of economy, as well as its thermodynamic superiority to the four-stroke engine, it costs less because unrefined oil is decidedly cheaper. But this engine thwarted the intellectual habits of the engineers of the time and it took 20 years to become widely used. It was criticized firstly for its weight, which is certainly greater per unit of horse-power than that of the four-stroke engine. It was also criticized for the noise it made when working and the particularly unpleasant smell of its exhaust. Diesel himself was one of the main reasons why industry took so long to adopt his engine: until his death in 1913 the engineer actually demanded that the engines built under licence fitted his rigorous specifications and, in particular, that they were designed to function at a constant temperature; Diesel wanted to keep strictly faithful to Carnot's theory. The problem was that his engine functioned much too slowly when kept at a constant temperature; in order to attain the desired output it had to be much more powerful, for it is chiefly at a low running speed that such an engine has most of the problems mentioned above. The diesel engine fulfilled its true potential when it was improved after its inventor's death.

Diesel had done a considerable service not only to mechanics but also to the four-stroke engine. Indeed, he made it clear that the output of a two- or four-stroke internal combustion engine was the result of the **rate of combustion** described above. When its combustion rate was raised the four-stroke engine produced a considerably improved output.

It is appropriate here to acknowledge two forerunners to Diesel: the Englishman James Joule who was the first in 1885 to try to design Carnot's ideal incomplete cycle, which he did by returning to a **porous piston** through which the exhaust escaped, and his compatriot Akroyd Stuart who made an **oil engine** (using a fuel similar to kerosene) which worked at low pressures.

Fuel cell

Bacon, 1959

A fuel cell is one where the reactions producing the electric current are obtained from substances contained outside and not inside the casing. Its advantage lies in providing a continuous current, which is not the case with classical reactors and electrolytic accumulators. The fuel cell seems to have been first designed in the mid-19th century, but only in theory, as attested for example by the Englishman William Henry Grove's **voltaic gas battery.**

It was however the Englishman Francis Bacon who built the first specific fuel cell in 1959. It contains an alkaline electrolyte (solution which conducts electricity), **potassium hydroxide**, dissolved in water. The electrodes are made out of a porous metal into which this electrolyte can only penetrate in a controlled way; behind one of the electrodes, which is in plate form, is oxygen and behind the other, hydrogen. When these enter into contact with the ions of the electrolyte, in the pores of the corresponding electrode, some electrons are freed, then captured on the other side by the atoms of oxygen. With the terminals of the electrodes being connected to an external circuit, the current flows for as long as there is hydrogen and oxygen in the reservoirs. It will continue to work without interruption provided that the reservoirs are refilled. Fuel reactors are used a great deal in **astronautics**.

Geothermal energy (exploitation of)
Larderel, 1818; Conti, 1903

It is difficult to assign a date to the first exploitation of geothermal energy and, for the same reason, to attribute to it an inventor. Steam or hot water shooting up from the depths of the Earth must surely have inspired the 'crude' use of geothermic energy in bygone days. In the 15th century the Florentine traveller Niccolo Zeno reported that in Greenland he had seen naturally heated greenhouses in which there were orange trees, apple trees and rose bushes; hot running water was distributed around the houses of the Nordic colonies in Greenland. In numerous regions of the world hot springs were used from very early times for rudimentary heating systems or for supplying the sometimes therapeutic public baths. The Frenchman Francois de Larderel was the first to exploit the geothermic emissions of steam; he did it in 1818 in a little village in Tuscany which to this day owes to him its name, Larderello. If Larderel was an inventor, he was one only *in partibus* for apparently he never planned a more elaborate exploitation of geothermic energy. It was Arago in 1848 who drilled the hot water wells of Grenelle.

The real pioneer in the subject seems to have been the prince, Giovanni Conti, who built a pilot station at Larderello for the transformation of geothermal energy into **electric energy**; it succeeded in lighting up four incandescent lamps. The Larderello power station eventually provided more than 290 megawatts (= millions of watts), which supplied the electric railway network of central Italy. More than half a century went by between the construction of this power station, which performed magnificently, and the construction of the next one in 1958 at Wairakei, New Zealand.

Geothermal energy can be either **high energy**, of which deposits are found in abnormal heating zones, or **low energy**, which exists in tepid aquiferous layers. In the former, the temperature reaches 200 to 300°C; these springs are very rare and there are only five known in the world, Larderello, the geysers in California, Valle Caldera in Mexico, and Matsukawa and Otake in Japan. In the last the temperature is between 60 and 90°C; they are found at depths of 1000–2000 m and are more common. Though occasionally exploited since ancient times as a form of heating, during the second half of the 20th century they have begun to be used much more; it is considered to be a largely underestimated source of energy, especially for heating; only high energy sources can be used to produce electricity.

Heat pump
Lord Kelvin, 1851

The principle of the heat pump dates back to nearly a century before it was first made. It was thought of by William Thomson, Lord Kelvin. It consists essentially of a refrigerator in reverse, which produces cold with the help of an evaporator and emits heat through a condenser. In the heat pump or **thermopump**, the heat produced by the evaporator is collected using a compressor. Thus it can be used for heating.

> The manufacture of reversible air conditioning devices began around the middle of the 20th century, thanks to the valves which could reverse the circulation of the air.

High-temperature superconductivity
Michel and Raveau, 1981; Bednorz and Müller, 1986

Superconductivity, which is the capacity of a body to allow an electric current to circulate without resistance, and so without energy loss, is an extremely complex phenomenon which was discovered in 1911 by the Dutchman Kamerlingh Onnes. It has both scientific and technological importance. Scientific, because superconductivity gives insight into the structure of matter and the behaviour of particles, which are a long way from being completely understood. For example, it remains to be explained why, in superconductivity, the electrons arrange themselves in pairs. Instead of repelling each other, which identically-charged particles usually do, they form closely joined couples (a phenomenon which can only be interpreted with the help of **quantum mechanics**). Technological, because the suppression of the electrical resistance in materials reduces the loss of electricity in transmission, and prevents heating (which, in the area of **semiconductors** would greatly improve performance). It also justifies the idea of storing electricity (that is, the possibility of batteries capable of supplying motor cars for long journeys), and the completion of very fast magnetic levitation trains.

From 1911 until 1986 superconductivity could only be achieved at very low temperatures, around **absolute zero**, −273°C. Although there were numerous scientists who predicted practical applications for it, their efforts to raise the temperature at which it worked were unsuccessful; in 1986 the highest temperature to be obtained was −250°C, or 23.3 Kelvins (1 **Kelvin** is defined as 1/273.16 of the thermodynamic temperature of the triple point of water, and degrees Kelvin = degrees Celsius +273.16) — which meant that superconductivity could only be obtained together with energy expenditure, which is necessary for cooling, and far above that which could be economical. However in 1981 Claude Michel and Bernard Raveau and their colleagues from the Crystallography

> John Bardeen, one of the inventors of the transistor (see p. 84), together with Leon Cooper and Robert Schreiffer, formulated the theory of super conductivity in 1957. The heart of this theory is the formation of pairs of interdependent electrons and the condensation of these pairs in a unique quantum state. It was Chakraverty who completed this theory with one suggesting that the transition of a material from a non-conductive to a conductive state is assured by two phenomena: the changing effect of temperature and the interaction between the electrons and the quanta of vibrational energy in the crystal lattice, or phonons.

and Material Science Laboratory of Caen University synthesized some metallic oxides — **copper** and **lanthanum oxides** doped with **barium** or **strontium** ($La_{2x}Ba_x CuO_{4-y}$, for example, — representing the quantity of lanthanum which is substituted for barium and y the amount of oxygen lacking). They proved to have the property of being excellent conductors in peculiar circumstances: the doping modified the ways the electrons passed through the crystalline mesh of the oxides, by altering the interaction of the electrons with the vibrational energy quanta of the crystal lattice, called **phonons**. This model provided confirmation of the theoretical explanatory model put forward in 1979 by Benoy Chakraverty, from the National Centre of Scientific Research in Grenoble.

Michel and Raveau did not research superconductivity, but their work is inseparable from the birth of (relatively) high-temperature superconductors. Using these experiments as a basis, in 1986 Georg Bednorz and Alex Müller, from the same IBM research laboratory in Zurich where the **scanning tunneling microscope** (see p. 157) was invented, studied the type

of oxides that Michel and Raveau had made, this time specifically from the point of view of superconductivity. They saw a considerable fall in resistance at temperatures of 23.3 K to more than 30 K. Following verification, that is, after having observed the three phenomena or 'signatures' specifically connected with superconductivity (disappearance of magnetism and the peculiar specific heat, as well as the lowering of resistance), they published their work in *Zeitschrift für Physik*. Bednorz and Müller were awarded the Nobel Prize for physics in 1987 (within a record time of a few months of their discovery). In the same year, 1986, the Japanese Koichi Kitazawa from Tokyo University, the American Paul Chu from the University of Houston, and then the American Laboratories AT&T-Bell at Murray Hill achieved superconductive effects at temperatures above 40 K. Then superconductivity at 52 K was recorded (under a pressure of 12 kilobars), and even between 80 and 93 K, that is, around −200°C. At the end of the 1980s the possibility of obtaining superconductivity at ambient (room) temperature seemed in sight.

Magnetohydrodynamic (MHD) generator
Branover and Solmecs, 1988

It is not easy to attribute a name and a date to the invention of the magneto-hydrodynamic generator which can essentially be defined as a classical generator which uses a **gas** or a **hot metal as conductor** instead of the classic copper coil of conventional generators. Michael Faraday had the grandiose plan during the 1830s of throwing iron filings into the Thames and installing giant magnets on the river banks. As water is a conductor of electricity the passage of iron filings in front of the magnets would be enough to charge it with electricity, without any energy expenditure; the electricity would be collected with the help of giant electrodes submerged in the water. Though a brilliant theory, it was

a flight of fancy on Faraday's part and would in fact have been rather difficult to put into action and make profitable.

Nevertheless, the replacement of copper wires by hot fluids has been the object of perfectly realistic research since the 1970s and it has given rise to a new domain which is called **magnetohydrodynamics** or **MHD**. The United States, France, and numerous other countries have pushed forward the theoretical and practical work on MHD for the main reason that it produces about 50% more energy than a classical generator using the same fuel. For example, in the heating of coal to some 2760°C, a temperature approximately double that reached by a classical generator, the Americans obtain a gaseous plasma into which they inject potassium salts to increase its electric conductivity, and then they re-inject this plasma into a magnetic tunnel, where it surges with the force and the speed of a jet of gas from a rocket nozzle, thus generating electricity. The advantage of the American technique is that the remaining plasma, though rather cooled down, can be used again, this time to make a classical generator function. This means a 50% efficiency of conversion into electricity, compared to the 35% of classical generators. In other words, the amount of coal required to produce 2 kWh (kilowatt hours) in a classical generator would produce 3 kWh in a MHD generator.

When in 1988 Americans and Soviets planned the completion of a commercial generator for the mid-1990s, the Israeli inventor Herman Branover, director of the MHD Centre at the Ben-Gourion University in Néguev in Israel, and the US–Israeli firm Solmecs Ltd had succeeded in making a prototype of a commercial generator using a mixture of molten lead and bismuth as the liquid. This had given satisfactory results and the commercialization was planned for 1990. This prototype, called Etgar 5, only converted 46% of the available energy in the fluid, but it could use coal, oil or schist-derived oil. Its characteristic is to work only at temperatures five times higher than those of the great American and Soviet prototypes, and so suits countries which do not have oil or petroleum resources at their disposal.

Semi-diesel

Price, 1914

Diesel's engine (see p. 91) posed a problem when it was first made, i.e. the need for a **high-pressure air compressor**, which itself had a high fuel consumption. Moreover, it had a defect in that the sudden expansion of the highly compressed air in a cylinder where the pressure was lower caused cooling which slowed down the ignition; in fact, any reduction in pressure caused the gas to cool. Diesel had used pressurized injection only to insert powdered coal into his cylinder but when liquid fuels replaced coal, the injector could be replaced by a **pump**. This did not solve all the problems, such as the very thick **exhaust fumes** from the engines. The combustion was as yet imperfect, even at running speeds proportional to the size of the engine.

It was suddenly realized that in a way the type of pump being used for liquid fuels was not effective: being arranged in battery at the rate of one pump per cylinder, in theory the fuel was injected at the most profitable moment in each cylinder cycle. But the pumps released the fuel in too weak a way to assure complete combustion. Each drop of fuel set out 'in search', so to speak, of oxygen molecules to burn; since oxygen represents only 20% of air, the combustion was far from perfect, hence the fumes. The injectors were then altered by creating some interior air currents which would offer the greatest opportunities of combustion to the fuel. The results were more or less satisfactory. In 1914 the English engineer William T. Price resolved to bypass this problem. Paradoxically, he lowered the pressure in the cylinders; in theory, combustion would be no longer possible. But Price achieved it by installing a Nichrome wire in the cylinder, which is a nickel–chromium alloy of very high resistivity. It also heats up rapidly under the effect of an electric current and becomes white hot. Combustion occurs in this way therefore, and much more thoroughly than with the previous engine. Price improved it once more, after he had registered his patent, by injecting part of the fuel at the end of the cycle in order to maintain the combustion of the Nichrome wire. This is what was called the 'semi-Diesel', because it was in two stages, with a compression rate nearly halved. It was also much more economical.

Servo-motor

Farcot, 1853

A servo-motor is the name for any mechanism playing an intermediary or final role in the working of a **regulating system**. The term was coined by its inventor, the Frenchman Jean Jacques Farcot. In 1853 he adapted a regulator to the steam machine which enabled it both to function continuously and have variations in rate; this is what he called an **isochronous regulator**, which means of equal time. This was however only the embryo of the servo-motor, for Farcot pushed still further his philosophy that the machine should serve man, so that automation can relieve the operator of constant surveillance and intervention. In 1863 he was the first to design a way of working the motor by directing the energy for it using manual controls. The first actual servo-motor was installed that same year on a ship, for use in steering with the tiller. The 'commands' were traditionally transmitted by chains or cables activated by a helmsman. The servo-motor made it possible for the commands to be transmitted by a rod which activated a distributor of pressurized fluid; this enabled the commands to be more subtle. There was a clever principle behind it: the fluid under pressure in the cylinder caused the piston to be displaced in one direction or another, with the injection of fluid taking place either above or below the piston. The moving of the 'distributor slide valve' controlled the movement of the piston and consequently that of the tiller by a system of connecting rods.

Szilard's engine

Szilard, 1929

The essential characteristic of a **heat engine**, such as a refrigerator, is that it produces a difference in temperature which requires the expenditure of energy to maintain it. It is an engine which obeys the **second law of thermodynamics** which rules that an isolated system can never pass through the same physical state twice. In other words, cold molecules cannot become hot molecules without the use of energy. This law or principle which postulates the **irreversibility of physical processes** has been the subject of countless studies since its formulation by the Frenchman Sadi Carnot and even today it feeds powerful philosophical speculations.

In 1871 the Scot James Clerk Maxwell proposed a theoretical device, now famous as **Maxwell's Demon**, which contravened the second law of thermodynamics. He imagined a creature that sat at a tiny door between two compartments of a container, one of which was empty and one which was filled with a gas. The demon would watch each gas molecule and let through only the fast molecules. In this way the demon would separate the hot molecules from the cold molecules, heating one compartment and cooling the other. Put into practice this scheme could provide a limitless supply of energy in contradiction to natural laws, in particular the principle of Carnot.

Scientists have long wrestled with the idea of Maxwell's Demon, trying to find a flaw in the concept and thus uphold the laws of thermodynamics.

In 1929 the famous Hungarian physicist Leo Szilard devised a resolution of the problem which in time became the basis of information theory and the quantitative basis of thermodynamics. Szilard invented an entirely theoretical engine. This consisted of a box with a piston at each end, containing just one molecule. The demon extracts energy by first detecting which side of the box the molecule is on and then dropping a partition and moving the piston into the empty half of the box up to the partition. When the partition is removed the molecule will push the piston back thus transferring energy to it. This contradicts the second law of thermodynamics.

Rudulf Landauer of IBM showed that although the Szilard engine gains energy it uses an equivalent amount of energy in the process of making the decision (about which side of the box the molecule is in).

This is the only drain in energy because storing the information needs an insignificantly small amount of energy and the piston can be frictionless. One bit (binary digit) of information is required to describe which side of the box the molecule is in and Landauer states that the energy gain is exactly that required to change that bit. Through this argument information content can be related to **entropy** in a thermodynamic system.

Everyday Life

The inventions which have changed our daily life since 1850 would seem too many to list. But, if the 'gadgets', which have not greatly changed our mode of living, are ruled out, only a few really important ones remain. These can be briefly listed as: the light bulb, a lasting wonder for which we can never be grateful enough to Edison; the credit card, which has resulted in a change in the financial structure of all industrial countries; detergents, an invention of rather ambiguous value to civilization, since it threatens to pollute our drinking-water supply; the automatic dishwasher, toilet paper and the rubbish bin, which have greatly improved our levels of hygiene, not to mention toilets themselves, and the ball-point pen, an instrument whose success has come, paradoxically, at a time when the art of writing is in decline.

Other inventions which affect modern daily life either came before 1850, such as the telephone, the typewriter and the lift, or are included in other sections of this book, for example television, radio, the car and synthetic textiles.

There are more less obvious facets of progress — mass production and miniaturization, which enable us to carry around a pocket calculator which is thinner than any wallet, and which 40 years ago would have filled a room. These contributory factors are at least as important as the finished product itself, and the observation is valid for many other inventions featured in this book. Without the miniaturization of electronics the reliability of the airlines which take us away on holiday would not be possible, and without the development of mass production, the headphones which are an essential part of personal stereos, would probably cost a great deal more.

Aerosol (spray)
Rotheim, 1926; Kahn, 1939; Goodhue and Sullivan, 1941

An aerosol consists of fine droplets of a liquid or a solid being suspended in a gas; these vary in size from 1 to 50 millionths of a millimetre. The first person to think of the dispersion of a liquid in this way was the Norwegian Erik Rotheim; he was inspired by the **vaporizers** on ordinary perfume bottles, in which the injection of pressurized air causes the diffusion of the aerosol through a tiny hole. This invention did not come about until 1939 when the American Julius S. Khan advocated injecting the pressurized gas into a disposable container. Its first use was effected by his compatriots

L. D. Goodhue and W. N. Sullivan who commercialized the first disposable insecticide aerosols in 1941.

> The gases used in the manufacture of aerosol sprays, **chlorofluorocarbons** (CFCs), are suspected of contributing to the destruction of the **ozone layer** which filters utlra-violet rays. In 1989 an international conference held in London discussed plans for the banning of all production and use of CFCs by the end of the century.

Automatic toaster
Strite, 1927

Obviously tired of eating burnt toast, the American mechanic Charles Strite designed a machine with a spring device controlled by **thermocontact** which ejected the bread at a given point during its cooking. This was the first automatic toaster, called **toastmaster**.

Ball-point pen
Biro, 1938

The ball-point pen was invented in 1938 by a Hungarian journalist, Laszlo Biro. Impressed by the quick-drying ink used in a paint shop, he made a prototype of a pen based on the same principle. He developed the idea in Argentina, where he had gone to escape the Nazis. He patented it in 1943 and his pens were first sold in Buenos Aires in 1945. In 1953, a French baron, Bich, developed a process for making ball-point pens which enabled them to be produced at a much lower price.

Breakfast cereals
Perkey, 1893; Kellogg, 1894; Kellogg, 1898

The first ready-to-eat breakfast cereal was Shredded Wheat, produced in Denver, Colorado, in 1893 by Henry D. Perkey. He had the idea for marketing it after being told by a sufferer from indigestion that he ate boiled wheat soaked in milk every morning, to soothe his stomach. At around the same time, William Kellog was experimenting at his brother's sanitarium in Battle Creek, Michigan, with different ways of preparing wheat to make a more digestible substitute for bread. After one batch of boiled wheat was allowed to stand, William Kellogg had the idea of flaking it. It was launched commercially as Granose Flakes in 1894. In 1898 Corn Flakes were first marketed by William Kellogg's Sanitas Food Company.

Carpet sweeper
Bissell, 1876

In 1876, before the vacuum cleaner had become a domestic appliance, the American Melville Bissell invented a broom which would sweep away dust without stirring it up in the air. The invention has an interesting origin: when Bissell worked in a porcelain packing factory he was persuaded (and undoubtedly quite correctly) that the lasting migraines from which he suffered were caused by the dust from the straw.

Chewing gum
Adams, 1870

Chewing gum is an American invention and was originally meant as a substitute to be used by tobacco-chewers at the end of the 19th century. Thomas Adams of Staten Island created it out of **chicle** (a gum-like substance obtained from the sapodilla, a tree which grows in the Yucatan peninsula of Mexico). By 1872 Adams had founded his first firm and commercialized his invention (which was a chance discovery in the search for a substitute for the latex of the hevea, from which rubber is made).

Credit card
Scheider, 1950

The first credit card, which was invented by the American Ralph Scheider in 1950, offered only very restricted credit. It enabled the cardholders, who were all members of a club, the Diner's Club, to eat in 27 New York restaurants and settle the bill later. When the Bank of America created the first **bank card** in 1958, Bankamericard, the services of the Diner's Club had already expanded considerably, as had the network of affiliated establishments; it subsequently became possible to pay by card for hotel bills and even for purchases in particular shops.

Four decades later the proliferation of different card systems and their networks has expanded in such a way that a new economic index has arisen from the amount of money in circulation due to this kind of credit.

Cremation

Le Moyne, Pini, 1876

Cremation is a very ancient practice for the dignified disposal of corpses. There is a description of it in the *Iliad*. It came to an end in the year AD 100 — save under exceptional circumstances, such as plague epidemics — under the influence of Christianity and the belief in the resurrection of the dead. When Queen Victoria's surgeon, Sir Henry Thompson, suggested in 1874 that it should be brought back into effect, by arguing that the decomposing corpses in the cemeteries were a danger to the health of the people, his opinion was met with a certain resistance. Although Thompson had created a Society for Cremation in England (to which some great writers belonged, such as Trollope, and members of the aristocracy, such as the Duke of Westminster), the lawmakers still took a long time to agree to it; it did not become legal until 1884. Several inventors set to work on a satisfactory cremation system, for they foresaw the technical problems of **incineration**, which could not take place in the open air as it had done in ancient times.

Independently and simultaneously in the year 1876 the American Julius Le Moyne and the Italian Pini, from the Cremation Society of Italy, invented the **crematorium furnace**, a fiery chamber inside which an intense heat takes about an hour to reduce the corpse to a small volume of ash.

Crossword

Wynne, 1913

A family game called the Magic Square had been in existence in Britain since Victorian times; it involved drawing up squares consisting of a small number of boxes. These contained horizontal rows of five or six words arranged in such a way that the letters also made sense vertically. This was a distant derivative of **acrostics**, which had been in practice since the 16th century and whose inventor is not known. But the first crosswords as we know them today, consisting of grids which have to be filled in using clues, are the invention of the Englishman, Arthur Wynne, who was originally from Liverpool but settled in New York. They appeared in the Sunday supplement of the *New York World* newspaper in 1913. They were to be extraordinarily successful as an adult pastime.

Darts game

Royal Flying Corps, 1914

This skilful pastime is derived unexpectedly from a military invention, that of weighted steel arrows which the Royal Flying Corps, later called the Royal Air Force, had invented in 1914 for the aerial bombardment (which claimed many victims) of the enemy lines. After the war, a game was made with the surplus arrows.

Detergents
Anon., 1916

The first synthetic detergent was invented under the duress of wartime shortages; it was Nikala, made in Germany in 1916, with the aim of economizing on the fats which were necessary for the manufacture of ordinary soap and which were in short supply. The name of the inventor has been impossible to find. It was made from a short-chain type of **sulphated alkylnaphthalene derivative**, which was produced by combining **propyl** or **butyl alcohol** with **naphthalenes**. As an actual detergent it was only mediocre since it required rubbing but it was a good softening agent and, once the war had ended, the formula was taken up again to be improved. The United States made some synthetic detergents which were marginally more effective but which nevertheless did not have any appreciable commercial success.

The first detergent to be successful was Lux soapflakes developed by the British firm Unilever in 1921. This formula had the double advantage over ordinary soap of not leaving residue on the washing and in the basin in which the washing had been done, and of not yellowing during ironing.

The progress made in household detergents paved the way to the development of shampoos.

Dishwasher
Cockran, 1889

In 1879 the American Mrs. W. A. Cockran from Shelbyville, Indiana, set to work on the project of an automatic machine for washing dishes. Only 10 years later, having experimented with several prototypes, she came up with a model which could be commercialized. Her machine was rather cumbersome, being powered by a steam motor. It does not seem to have been very popular, and even when the **electric dishwashers** were made in 1906, also in the United States, their handling as well as their water leakage obviously meant that they were not very reliable machines. The production of the detergent Calgon in 1932 improved the results appreciably, but it was only in 1940 with the production (also in the United States of America) of a properly automatic machine that this domestic appliance became widely used. Its conquest of households all over the world dates from the 1960s.

Disposable paper handkerchiefs
Kimberley-Clark Co, 1924

These paper handkerchiefs were produced commercially for the first time by the American firm Kimberley-Clark and marketed under the name of Celluwipes. This was later changed to Kleenex. They were meant to prevent the spread of germs and facilitate the work of the housewife, as well as make working in laundries more healthy.

Ecology
Haeckel, 1866

It is difficult to think of ecology as an invention because some authors think it originated with Aristotle. The term, however, was coined by the German Ernst Haeckel in 1866; he defined the concept of a science which studies the **relationships between living beings and between living beings and their environment**. Indeed, as shown by the Frenchman Pascal Acot, the ecologist is born from the amalgamation of several disciplines, such as Humboldt's geography, Linnaeus's natural eco-nomy, palaeobotany and plant physiology. During the second half of the 20th century ecology has caused some changes in opinion which in some countries has led to the formation of political parties such as the Green Party in the UK. On the whole it has resulted in a great deal of trouble being taken to preserve and re-establish the balances in nature, by the re-introduction of nearly extinct animal species, and the protection of endangered plant and animal species by the banning of certain products.

Electric chair
Brown and Kenneally, 1888

This method of execution was designed by the Americans Harold P. Brown and E. A. Kenneally, who was the last director of Edison's team of electricians. It was thought to be conclusive after several experiments on animals and was used for the first time on 6 August 1890 to put to death the murderer William Kemmler in New York State Prison. It caused great controversy at the time: Kemmler took eight minutes to die. By 1915 it had been adopted by 25 US states.

Electric guitar
Rickenbacker, Barth and Beauchamp, 1931

In 1931 the Californians, Adolph Rickenbacker, Barth and Beauchamp had the idea of making some alterations to an ordinary guitar by placing a microphone on the back of the instrument and connecting this to a loud speaker. The resulting hybrid instrument is the ancestor of electric guitars. Soon afterwards the Electro String Instrument Corp. of Los Angeles produced a **banjo** made of steel and aluminium which was nicknamed 'the frying pan' and was subsequently a great success.

Electric hand drill
Fein, 1895; Black and Decker, 1917

The first electric hand drill was invented by Wilhelm Fein of Stuttgart, Germany in 1895, but the first with a trigger on/off switch was made by two American inventors S. Duncan Black and Alonso G. Decker. Their version weighed 11 kg; lighter and improved models were produced after World War II.

Electric heater

Burton, 1887; Downing and Crompton, 1892

In 1892 Dr W. Leigh Burton patented an electric heating system which was produced for sale in 1898 by Burton Electric Co. of Richmond, Virginia. The heater consisted of a cast iron case enclosing **electric coils covered in clay**; it was shaped like a table. The first buildings to be heated by such devices were the motor-stations of the Aspen Mining Company, Aspen, Colorado.

In the UK, in 1894 Crompton and Co, Chelmsford began producing heaters which had been designed by H. J. Downing. In these a wire was attached to a cast iron plate and protected by a layer of enamel. The Vaudeville Theatre Company, London, had a complete system of Crompton heaters installed in 1885.

The design for electric fires was improved by the American, Albert Marsh, who invented an alloy of nickel and chrome which could be heated red without its melting. In 1912 the Englishman, C. R. Belling made a reflecting clay around which such a nickel-chrome wire could be wound, and he produced the first electric bar heater in the same year.

Electric light bulb

Staite, 1845 and 1850; Draper, 1846; Shepherd, 1850; Farmer, 1858; Swan, 1865–70; Lodyguine, 1868; Edison, 1878–9

The electric light bulb was developed progressively from the **arc lamp**, starting with the observation made by the Frenchman Jobart in 1838 of the way in which a carbon rod sealed in a vacuum will light up on the application of an electric current.

In 1845 the American W. E. Staite had patented an **incandescent lamp** of this type. The following year the Englishman J. W. Draper made a **platinum filament lamp**. In 1850 Staite made another, as did the Englishman E. C. Shepherd. In 1858 the American Moses G. Farmer invented automatically-controlled platinum lamps, as a result of which he could light up a room in his house for several months at a time. In 1868 the Russian Lodyguine lit up the St. Petersburg harbour using 200 lights which he had invented. The Englishman Joseph Swan had himself been studying various types of electric light since 1865.

Many others across the world carried out research on the use of electricity for lighting; among them there was another Russian who must be mentioned, Paul Nicolaïevitch Jablochkoff, who made arc lamps under glass. These were quite effective for they were used for lighting the Moscow–Koursk railway and several public and private houses as well as the streets of St. Petersburg. Every one of these precursors came up against the same problem, which was both scientific and technological: scientific because no one really knew what effect the electricity had on the carbon, and technological because it was difficult to make a carbon thread fine enough and to put it in a vacuum without breaking. This was the reason why electric lighting was proving to be no competition for gas lighting at the time when Edison tackled the problem in 1877; even though the arc lamps illuminated some places in Europe, such as the Avenue de L'Opera in Paris, the large bulbs only lasted about two hours.

Edison had little more knowledge than his predecessors but he did have the virtue of tenacity: he successively tried a great number of different materials in order to make his filament and tirelessly endeavoured to perfect the **vacuum** inside the bulb. During the autumn of 1878 he thought he had found the solution with a platinum filament, but this was just a repeat of Farmer's experiment. He patented it, but it was no more durable than the others. Edison tested other metals, but at the end of January 1879, after still many more attempts, he abandoned the idea of metal filaments.

The particular problem of the vacuum in the bulb and of how to seal it still remained. It began to be solved during the autumn of 1879 when Edison finally produced pumps which were powerful enough to achieve a vacuum of a millionth of standard atmospheric pressure. On the 21 and 22 October 1879 he renewed his experiments with a bulb, the carbon filament of which had been lodged — not without difficulty — in a groove of nickel wire, and in which he had achieved the indispensable vacuum; for the first time he obtained a light which lasted for 40 hours. This was still only a short time but he was on the right track — a fortnight later he submitted a request for the patent. Innumerable experiments with the shape of the bulb, how it was put together and the thickness of the filament were to take place before Edison managed to make a bulb which had a satisfactory life span. It was not until the end of 1880 that he made a 16-watt bulb which lasted for 1589 hours.

He continued with his research on other materials which could provide a better filament, from bamboo fibres to coconut hairs to fishing line, while the bulb factory which he had founded began to function. By the end of 1882 it was producing 100 000 bulbs each year, and in 1892, 4 000 000. Despite strong opposition in scientific circles to Edison's invention, networks for the distribution of electric current began to grow up everywhere.

Edison's invention, which attracted a large audience at Menlo Park, for he also had an acute sense of business and publicity, generated a formidable wave of hostility from scientific circles. The respectable magazine *Nature* ridiculed his plans and the figures which he had put forward on the cost of electric lighting, contesting among other things the price of one shilling which Edison had indicated. The city of Fort Myers in Florida, where Edison had a home, refused his offer to light the city streets by electricity at his own expense. Nevertheless one of the very first American citizens to light his New York home using Edison's lamps and a **current distribution circuit** was the well-known banker John Pierpont Morgan.

Electric mixer

Universal, 1919

The growing electrification of domestic appliances led the American firm Universal Co. to make a mixer in 1919 which could be powered by mains electricity. The enormous success of this machine has not diminished.

Electric organ

Cahill and Hope-Jones, 1895; Couplex and Givelet, 1930; Hammond, 1935

The electrically-controlled organ was invented in 1895 by both the American Thomas Cahill and by the Englishman Robert Hope-Jones independently; it had been tested experimentally in 1860 and in 1888 the Englishman Henry Willis, an organ-maker, had already made a proto-type which was considered satisfactory by Canterbury Cathedral. But it was Cahill and Hope-Jones who almost simultaneously invented the following principle: the pressure applied to a key closes an electric circuit; this activates an electromagnet, which finally activates a **pneumatic pump**; the air enters the **pipe** and the same motor controls the input of air to the pipe by the **sliding stop**. It should be noted that this invention, which, unlike the traditional organ, has to be tuned only every 50 years, is not exactly similar to the actual electronic organ, which was invented by the Frenchmen Coupleux and Givelet in 1930, but it nevertheless marks the application of electricity to the production of musical sound in one of the oldest instruments in the world

(see *Great Inventions Through History*). This invention was to offer much flexibility in the production of the sounds. It was capable of playing **chords**, but the **interference** which was produced at the outset, and which meant that tuning was required, could only be gradually eliminated and it was not until 1943 that they were completely suppressed by the Frenchman Constant Martin. The electric organ was to culminate in the creation of the **liturgical organ**, which imitated the pipe organ. In 1935, the American John Hammond thought of producing the oscillations of an electric current using **rotating electromagnetic generators** and made an instrument which was to be known under the name of the **theatre organ**.

Electric oven
Bernina Hotel, 1889

The first electric oven was the one invented in the Bernina Hotel, Samaden, Switzerland. This hotel actually had an independent electricity supply from a dynamo which worked from a nearby water-fall, and so an oven with a heating element was built which could be used, among other things, for the baking of bread and confectionery. The following year the first electric oven for public sale appeared on the market, made by the firm Carpenter Electric Heating

Manufacturing Co., of St. Paul, Minnesota, in the United States.

The first electric oven manufactured in the UK was designed by H. J. Dowsing for Crompton & Co. (see **Electric heater**); it was based on the Carpenter model. Electricity was a very expensive form of energy at that time, and electric cookers did not become economically competitive with gas until charges began to be reduced in 1906.

Electric razor
Anon, USA, 1900; Schick, 1928

Several different electric razors were patented in the United States in 1900. They consisted of instruments based on the principle of a **revolving cutting head** which cut the hairs through a **grid**. Apparently

they were not satisfactory, for the first electric razor to come into reasonable widespread use was the one patented in 1928 by Jacob Schick.

Electric toothbrush

Scott, 1885

This electric version of the ordinary dental hygiene accessory seems to date from the 1960s, the time when it became commercially successful. But in fact the first patent on a vibrating toothbrush with an electric motor dates back to 1885 and was submitted by the American Scott (Christian name unknown). The device was inconvenient, noisy and expensive, which explains why it did not meet with the enthusiasm its inventor undoubtedly expected.

Espresso machine

Gaggia, 1946

The growing popularity of coffee consumption led the Italian firm Gaggia in Milan to produce a **steam machine** which could prepare concentrated coffee quickly by passing steam through a filter; the original process of making beverages using steam, called **infusion**, which is different from **decoction** (extraction by boiling) was very old, but slow. The new machine was to popularize the taste of coffee which was strongly flavoured and, contrary to common opinion, low in **caffeine**.

Extinguisher

Carlier, 1866

When in 1866 the French doctor Francois Carlier mixed together some **bicarbonate of soda** and **sulphuric acid** he obtained a composite which he claimed could be used to extinguish fire. Carlier thus made the first device to extinguish fire which used a product other than water. In 1905 the Russian Alexander Laurent changed Carlier's formula by replacing the sulphuric acid with **aluminium sulphate**, which had the same effect: the formation of a layer which separated the burning material, such as petrol for example, from the air, provided that the material floated. Since then the products inside extinguishers have been improved further, to match the type of substance which they are meant to extinguish. Consequently, for grease fires **halogenated hydrocarbons, foams** and **powders** are used.

Food processor

Wood, 1947

This kitchen appliance was designed by Kenneth Wood to perform a number of different tasks. It consisted of a sturdy and powerful engine component to which accessories could be fixed: mixer, mincer, slicer, can opener etc. It was marketed as the Kenwood, and many other models are now supplied to the domestic market.

Frozen food

Birdseye, 1923

The inventor of commercialized frozen food was the American leather merchant Clarence Birdseye, who founded an industrial plant for the production of uncooked frozen foodstuffs in Gloucester, Massachusetts, in 1923. In 1930 his firm put frozen peas on the market (this was the only food with which he dealt). **Pre-cooked frozen food** did not appear until 1939 and were introduced by Findus, one of Birdseye's competitors.

Gas chamber execution

Turner, 1924

One of the most sinister inventions connected with capital punishment was that of the American Major D. A. Turner from the US Army Corps. It was used for the first time in 1924 for the execution of the murderer Gee Jon in the Nevada State Prison in Carson City. The condemned man took 10 minutes to die.

Jeans

Davis, 1870; Levi-Strauss, 1872

If the definition of the trousers known under the name of 'jeans' is accepted to refer to those made of blue twill with pockets attached using metal rivets, then they were first made by two American expatriates, Jacob W. Davis from California who made the **riveting**, and Levi-Strauss, the clothing manufacturer with whom they were patented.

In the *Oxford English Dictionary* the term **'jeans'** is said to originate from the word **'Genoese'**, which refers to a woven fabric made in Gênes; the etymology is possible, but does not exclude the fact that the word could also be derived from the spinning machine which had several spindles, the **Spinning Jenny**, made by the Englishman James Hargreaves in 1785. Nevertheless it is fact that the working clothes which were the forerunners to jeans, which were then called **'denims'** and were made of **serge**, were made out of this material from the 18th century onwards and it was imported from Nîmes. Denim was used by Levi-Strauss from 1860, before the jeans were 'invented'.

Launderette

Cantrell, 1934

It was the American J. F. Cantrell from Fort Worth, Texas, who had the idea in 1934 of an establishment where one could use a washing machine and do one's laundry oneself, in return for payment by the hour. The first launderette was called a *Washeteria*.

Lego
Christiansen, 1955

Lego bricks are small multicoloured bricks made of plastic which fit together so that stable models and structures can be made. They are a standard part of every toybox. They were designed by a Dane, Ole Kirk Christiansen, a carpenter who had retrained as a toymaker. The name Lego is derived from the Danish *leg godt* ('to play well').

Light organ
Scriabin, 1908

The idea of an organ with a keyboard which would control not the production of sounds but that of the luminous flux of different colours was conceived by the Russian composer Alexander Scriabin in 1908 with the aim of using it to accompany the playing of his orchestral poem *Prometheus*. The poem was played for the first time in 1911 without this unheard-of instrument, which seems only to have appeared in 1922, under the name of 'Clavilux', promoted by the composer of Danish origin, Thomas Wilfred, who had a similar goal. Wilfred, to whom the invention is incorrectly attributed, had the playing of his *Triangular Study* accompanied by a light organ.

Meccano
Hornby, 1901

The first **building game** consisting of compatible metal pieces, meant to stimulate a child's creative spirit and taste for mechanics, was invented in 1901 by the British game designer Frank Hornby.

Microwave oven
Spencer, 1945

The microwave oven applies the principles of radar to cooking food (see **Klystron**). The idea was patented by Percy LeBaron Spencer in 1945 while he was working for Raytheon, a manufacturer of radar equipment. Raytheon manufactured the first oven of this type, the Radaraye, with a power of 1600 watts. This was a cumbersome and expensive machine, and its use was at first limited to institutional kitchens and hotels. The first hotel in the UK to install a microwave oven was the Kew Gardens Hotel, in London; it used a model manufactured in the UK, designed by R. J. Constable of the firm Lewispoint Holdings, called the Arctic Cooker. The first domestic microwave was marketed in 1967 by a subsidiary of Raytheon. Modern microwave ovens can be very compact machines, with a power output of 500 watts, or rather bigger ones with an output of up to 860 watts.

Monopoly

Darrow, 1933

The board game *Monopoly*, in which the players seek to acquire property and therefore rents, was invented in 1933 by Clarence Darrow, the lawyer, orator, pamphleteer and writer. He developed it while he was unemployed following the stockmarket crash of 1929.

Musical synthesizer

Moog, 1964

The electronic musical synthesizer, an instrument for making sound from **analogue electrical signals**, was designed by the American Robert Moog in collaboration with composers Harry Deutsch and Walter Carlos, in 1954. Moog gave his invention the name **synthesizer**, and the first one was produced for sale in 1964. It could produce only one sound at a time; since 1975 polyphonic ones have been developed. The synthesizer is not to be confused with the **electric organ** (see p. 107).

Neon tube

Claude, 1910–38

At the end of the 19th century, numerous physicists were experimenting with the production of radiation by the creation of electric arc dischargers between two electrodes. This was done using tubes where a vacuum had been made, but admittedly a certain amount of various different gases nevertheless remained. The Frenchman Georges Claude carried out such experiments using neon and in 1909 he discovered that this gas became fluorescent. In 1910 he drew attention to the possibility of lighting public places using neon tubes. It was not until 1936 however that Claude actually managed to complete the neon tube, which became very successful in 1940 in industrial lighting and in the manufacture of luminous signs. The light given out by these tubes was unstable, but the first improvement to the neon tube was made only in 1948, with the discovery of the **fluorescent powders** made of calcium and strontium halophosphates. These made the emitted light noticeably more stable. New progress was made in 1972 with the discovery of blue and green magnesium aluminate fluorescent powders, which together with red yttrium oxide have improved both the quality of the light and the **colour speed** (avoidance of colour distortions).

Non-stick pan
Grégoire, 1954

This is an example of an invention being happened upon in the course of a search for something else. Marc Grégoire was trying to perfect his fishing rods when he discovered the process which makes it possible to apply a **Teflon** coating to metal. (Teflon had been discovered by Roy J. Plunkett, an engineer at Du Pont in the USA, in 1938.) He founded the Tefal company in 1956 to make frying pans and sauce pans. Teflon is the commercial name for polyfluorstetraethylene, or PTFE.

Paper clip
Vaaler, 1900

One of the most mundane and most useful inventions, the paper clip, a metallic clip that attaches sheets of paper together, was invented in 1900 by the Norwegian Johaan Vaaler and patented by him in Germany.

Parking meter
Magee, 1932

The American Carlton C. Magee was a journalist, chief writer of the *Oklahoma City Newspaper*, and also president of the Businessmen's Traffic Committee. People were beginning to complain that they could not find convenient parking for their cars, and so he thought of installing along the pavements a new device which would have a hand-operated spring system controlling a needle, the movement of which would indicate on the dial the amount of time elapsed since the insertion of a coin. Three years later his Dual Parking Meter Co. began to manufacture this machine whose use was intended to reduce the congestion in urban streets.

Pressure cooker
Lescure brothers, 1954

The pressurized cooking pot, or pressure cooker is a pan with a cover which locks into place to form an airtight seal. The steam produced during cooking is trapped inside; as the pressure builds up, foods are cooked at temperatures higher than those which can be obtained in a standard saucepan. The pan is provided with a **safety valve** to allow the pressure to be vented before it can become too high. It was invented in 1954 by Frédéric, Jean and Henri Lescure on behalf of the firm SEB.

Revolving door

Kannel, 1888

The invention of revolving doors was given its incentive by a phenomenon which appeared when the first sky-scrapers were built: the difference between the internal and external air pressure. When the lifts went up, they caused a suction effect which made the doors more difficult to open towards the outside. On the other hand, when the lifts came down, the doors tended to open themselves. In 1888 the American Theophilus von Kannel designed a revolving door which in theory was to solve the problem.

Rubik cube

Rubik, 1979

The Rubik cube was created in 1979 by the Hungarian Ernö Rubik. It was a complex version of the Diabolo of the 1930s, consisting of a large cube composed of twenty-seven small cubes, each outward-facing surface of which was a different colour, which had to be arranged in such a way as to make the large cube have monochromatic surfaces. The Rubik cube's originality lies in its system of **'universal' ball-and-socket joints**, which allow movements of 360°.

Safety razor

Gillette, 1895

The first razor with double edged blades called the 'safety' razor was patented in 1895 by the American King Camp Gillette. He marketed the invention through the Gillette Safety Razor Company which he founded in 1901.

Self-heating flask

Pozel, 1978

All one has to do is unscrew the top of the flask, wait for a minute, and the drink is heated by the effect of the heat produced by chemical reaction in a coil at the base of the container. The reaction is started by the rotation of the cap; this starts the ignition of the gas in the coil. The invention is by the Swiss firm Pozel, of Fribourg.

Self-service store
Alpha Beta Food, Ward, 1912

It was in 1912 that for the first time two grocery stores decided to economize on their number of assistants by allowing the customers to help themselves to the products which could be paid for at a cash till situated at the exit. These groceries were the Alpha Beta Food Market and Ward's Grocetaria in California.

Single-price shops
Woolworth, 1879

The idea of a shop where all the goods would be sold at a uniform price goes back to the American Frank Woolworth, the founder of the famous chain which bears his name. It was in 1879 that Woolworth opened the first shop where all the products were priced at either 5 or 10 cents.

Skateboard
Munoz and Edwards, 1968

The skateboard was invented in 1968 by the Californian surfers Mickey Munoz and Phil Edwards. At first it was called the **rollsurf** . . . and it was not very successful until around 1973.

Sun-tan cream
Schueller, 1936

Sun screen for the skin is actually a re-invention; since early times plant extracts have been used to activate the skin's natural pigmentation processes which reduce the harmful effects of ultraviolet as the **melanin** contained in the natural pigments serves as a screen. The plants used were citrus fruits, celeriac, angelica and parsley, being rich in **psoralens**, which stimulate the formation of melanin. These products were gradually forgotten as popular aesthetics in industrial countries tended to favour pale skin colour. At the end of the 19th century it was the discovery of soluble filtering substances, such as vegetable oils, which set cosmetology on the path to the commercial mass production of sun filters. With the **sun-tan fashion** having been launched by Coco Chanel in 1925, leading to some medical problems resulting from prolonged exposure to the sun, the need was for products which were easy to apply. The first one was Ambre Solaire, made in 1936 by Eugène Schueller, the founder of the French firm L'Oréal.

Swimming fins (flippers)

Corlieu, 1927

The first swimming fins, originally designed for **frogmen**, were invented and made out of rubber by the Frenchman Louis de Corlieu in 1927.

Table tennis

Gibb, 1890

Table tennis, or ping-pong, and its accompanying equipment was invented in 1890 by the British engineer James Gibb and produced industrially nine years later by another British firm, John Jaques & Son Ltd., with the London firm Hamley Brothers as distributor. Though it was named 'Gossima' at first, at the beginning of the century table tennis found its international — and alliterative — name of ping-pong. The first association of ping-pong players was formed in 1902, but it was only in the 1920s that this sport began to be internationally successful, and it was not until 1926 that the first world championships took place. Until the beginning of the 1960s, when the titles began to be more and more frequently won by Asian players, the large majority of champions were of Czechoslovakian and Hungarian origin.

Tea bags

Krieger, 1919

The packaging of tea in small muslin bags was first thought of and put into operation in 1919 by the American, Joseph Krieger.

Teddy bear

Mitchom, Steiff, 1902

The teddy bear, one of the most popular toys in the world, really is an invention. It appeared in 1902 simultaneously in the United States, where it was mass-produced by Morris Mitchom, and in Germany, where it was made by Richard Steiff. The invention has obscure origins. In the United States it developed from the desire to make a mascot which looked like the president Theodore Roosevelt, nicknamed 'Teddy', hence its name. In Germany, it was derived from the fact that the bear is the symbol of the city of Berlin.

Telescopic umbrella
Haupt, 1929

The idea of an umbrella that can be folded by sliding it into a small bag dates from 1929; it was first thought of by Hans Haupt, a German from Berlin.

Tennis
Wingfield, 1873

The modern game of tennis is an adaptation of the ancient French game of *jeu de paume*, which seems to have appeared in the 12th century, in which the expression '*Tenez!*', was used by the server before he hit the ball. As 'real tennis' the game was popular in England — mentioned in Shakespeare's *Henry V* — and a favourite pastime of Henry VIII. The rules of the game were drawn up in 1873 by the Englishman Walter G. Wingfield. He patented the game in 1874, but called it at that time *Sphairistike*, a Greek word for 'ball game'. In 1875 the All-England Croquet Club set aside one of its lawns for the game, which proved so popular that the club changed its name to the All-England Croquet and Tennis Club in 1876. The first Tennis Championships was held at Wimbledon one year later.

Time-keeping clock
Cooper, 1894

Until 1894, managers of firms had no means of checking the arrival and departure times of their employees other than having an official stationed at the entrance to the offices. But in this year the American Daniel M. Cooper devised a system consisting of a timekeeping machine synchronized with a clock which printed the time on a card when it was inserted. The card was divided into as many sections as the office had working days in the week (six at that time).

Toilet paper
Cayetty, 1857

A type of paper especially designed for use appeared in 1857, developed by the American Joseph Cayetty. It was considered for a long time in the rest of the world as a luxury item, a decadent form of refinement, and it only came into widespread use during the years following World War I.

Toilets

Crapper, 1886

One of the most disconcerting aspects in the history of humanity is the time it took the hygienists, engineers and public authorities to try to resolve a basic daily problem, the disposal of human excrement. At the rate of about $1^1/_2$ litres of urine and 150 grammes of solid waste a day, this problem had considerable importance for the town councillors.

It is thought that the Romans used a **chamber pot (matula)** for this purpose in the 3rd or 2nd centuries BC; it remained a basic domestic utensil for the next 22 centuries. The building of closets with fixed chairs with holes in them, such as those used by the Cretans during the 2nd millennium BC seems to have posed complex problems of architectural infrastructure and organization, although the Romans utilized a similar system in many of their military installations. Until the 18th century the chamber pot constituted the essential receptacle for excrement, discounting of course the corners and public places where people relieved themselves. These receptacles were emptied into the street, usually through the window, though later on and in certain areas into the gutter, so as not to soil the front of the buildings. Consequently the streets and the courtyards became fouled.

The aristocrat lived in the same condition as the peasant, relieving himself in street corners or by hedges. Some castles (and other buildings where large communities gathered, such as abbeys) had arrangements for disposal of waste in the form of special channels; sometimes a corbelled construction on the wall allowed natural relief to be directed into the open air — towards a ditch or moat, or, as in the case at Dunnottar Castle, south of Aberdeen directly over the cliff.

The unpleasant aspects of the lack of sanitation were endured as inevitable; they are chronicled by Pepys in his diary for 30 April 1666: 'at night home and up to the leads; but were, contrary to expectations, driven down again with a stink, by Sir W Pen's emptying of a shitten pot in their house of office close by; which doth trouble me, for fear it do hereafter annoy me'.

Perhaps such considerations inspired the Elizabethan poet Sir John Harington, to design a water closet, which was installed at his home at Kelston, near Bath in 1589. The invention was described in *The Metamorphosis of Ajax* published in 1596. It resembled a modern toilet in essential ways, such as having a reservoir of water constantly in the bowl, and a flush of water from a cistern to wash the contents of the bowl into a cesspit beneath. A water closet to his design was installed at Richmond Palace by Queen Elizabeth I (Harington's godmother).

The invention did not have great success at that time. Waste continued to be disposed of in public middens and into the rivers. Public toilets, where they existed, might be built over the river itself, as was the one over the Thames on London Bridge, which collapsed in the 14th century, plunging the user into the water below.

Sufficient numbers of people were persuaded of the desirability of a water closet to make a good living for London cabinet maker Joseph Bramah who patented his model in 1778. Between then and 1797 he sold 6000 water closets, which were designed so that the toilet bowl was boxed around by a wooden case. The front pedestal model was Jennings' Pedestal Vase, exhibited in 1884. This firm also introduced the oval seat. Before that the firm of Jennings had been the contractor for the construction of the first municipal public toilets (with water closets), installed outside the Royal Exchange in 1865. The charge for the use of this **public convenience** was 1d (one old penny) — a price which remained constant for most public toilets until 1971 (and the introduction of decimal currency).

The chief impediment to sanitary living was the lack of any organized sewage system. Septic tanks were some improvement on the basic bucket in the privy, but an effective one containing a **bacterial bed** which could decompose the organic matter by anaerobic **fermentation** was not invented until 1896, by Donald Cameron.

Before that, some important advances had been made. Underground sewers were laid in Hamburg, Germany for example in 1843. More important than that was Thomas Crapper's improved model of a flushing toilet, invented in 1886. He placed a tank containing water at a good height above the bowl, so that when the contents were released by the action of a chain on a lever system the flow of water had an expelling action, as well as a diluting one.

The dilution of waste was a great advantage as far as the rivers which received it were concerned. Crapper had also altered the design of the bowl, providing a **U-bend** so that clean water filled the pipe leading to the sewer pipe.

However, the invention could not be put to general use until the great Victorian effort in sewer-laying, and the supply of running water to all homes. Even after World War II, many homes in more rural areas of the UK were without flush toilets.

Traveller's cheques
American Express, 1891

The travel agency American Express invented the traveller's cheque in 1891. It consists of an adaptation of the old **letter of credit** which enabled the traveller to carry money with him which could be cashed in a foreign subsidiary company, without running the risks of loss or theft.

Vacuum cleaner
McGaffey, 1871; Booth, 1901

The principle of the domestic vacuum cleaner was sketched out in 1871 by the American Ives W. McGaffey; it consisted of a pump working an inverted propeller which in turn was powered by a **steam engine**. This was for industrial use. The next stage was reached by Hubert Cecil Booth in the UK, who put an **electric motor** in a wagon pulled by an animal; the dust in buildings could be sucked up using a long tube. His Vacuum Cleaning Company provided a cleaning service. It was not until 1907 that the American J. Murray Spangler, a night watchman, developed a much lighter machine and sold the rights to William H. Hoover, a harness maker. Hoover launched the Hoover Model 'O', a domestic cleaner which was powered by a small electric motor.

Velcro®
de Mertral, 1957

This method of fastening together objects, especially articles of clothing, was first conceived in 1948 by the Swiss engineer Georges de Mertral when he examined burdock seed heads under a microscope, and saw the minute hooks which enabled the seeds to cling to animals' coats and to clothing. It took eight years to develop the final product which consists of two nylon strips: one is covered with thousands of small hooks, the other with small loops. Velcro is so named from a combination of the French words *velours* ('velvet') and *crochet* ('hook').

Vending machine
Everitt, 1882

Machines which sell goods automatically function on the principle of the weight of a coin setting a release mechanism in operation. They seem a part of the commercial network of the 20th century, but in fact the first patent for a commercial machine of this type was submitted in 1883 by the Englishman Percival Everitt. It was only the institution of large scale networks which encouraged the spread of vending machines from the 1930s, beginning in the United States. Until then all selling had been done from shops or over the counter.

Video games
Bushnell, 1972

The first ever video game was the Pong, invented and commercialized by the American Norman Bushnell in 1972. It consisted of a device which worked on the principle of the **liquid crystal screen**: a knob at the side was used to direct a white spot towards a moving target. Encouraged by the success of his invention, Bushnell founded the firm Atari. Following a slight easing-off during the mid-1980s, due to the overproduction of this type of game and particularly of cassettes for home computers, demand for video games began to increase again in 1988.

Waterproof watch
Rolex, 1926

The first waterproof watch in the world was developed in 1926 by Hans Wilsdorf at the Swiss watchmaking firm Rolex. It was called 'Oyster' and was to prove invaluable to all those called to work in the marine environment. In 1927 Wilsdorf gave an Oyster watch to Mercedes Gleitz, who swam the Channel with it strapped to her wrist.

Zip-fastener
Judson, 1893

The principle of a fastener consisting of two rows of hooks which would be joined together by a slider first appeared at the Chicago Exhibition in 1893, presented by the American Whitcomb L. Judson. It was improved in 1912 in the United States by the Swede Gideon Sundback and in Europe by the Dane Catharina Kuhn-Moos. The system was used on a large scale in the clothing industry from the 1920s onwards.

Industry and Industrial Technology

The fact that the expansion of modern industry began in the second half of the 19th century would suggest that the inventions since then must have increased in a geometrical or exponential way. This is not the case at all, however, and it is actually only half a century later that the major inventions started coming to life. Stainless steel, for example, dates only from 1913, and butyl rubber from 1937.

Industry is slow at first to absorb chemical inventions, for which it later becomes terribly greedy.

Until the latter half of the 19th century, industry was essentially 'heavy'. It was only in the second half of the 20th century that really original inventions began to appear, such as memory alloys, kevlar, ferrofluids, and grafted textiles, the potential of which could still hardly be grasped even at the end of the 1980s.

In reality, this peculiarity can be explained by two chief phenomena. The first is automation, which has taken over the heavy industry since the mid-20th century and has enabled the same products to be produced at lower and lower prices. Accordingly, this has delayed the exploitation of new materials and processes, which are inevitably expensive. The second is the diversification, or rather the explosion, of that which was previously called 'industry' and which referred essentially to the iron and steel industry, now shrunk considerably. Original 'industry' has now more or less blended with the new giant electronics industry. Similarly, the car industry, which consitutes one of the last jewels of industry in the old sense, is now using many chemistry-based creations, such as metallic ceramics. Part chemical and part electronic, the car hardly remains 'industrial'. All the inventions which refer to it are to be found in the chapter on transport.

Even the term 'industry' is being slowly stripped of meaning. Thus nowadays the 'industry of biotechnology' refers to a domain where the equipment takes up less than a few hundred square metres. This is mostly in laboratories where the modern slaves, bacteria, are put to work in tubes to produce substances such as insulin.

Bactericidal paint
Henrion and SLVP, c.1970

This is a paint which contains substances which neutralize the bacteria and fungi responsible for **occupational diseases**; it was invented during the 1970s by the Frenchman Lucien Henrion, the founder of the Lorraine paint and varnish firm. The required types of paint are made to order.

Carbon fibres
Courtaulds Ltd., 1964

One of the most important inventions in the field of **synthetic fibres** must surely be that of carbon fibre. It does not actually belong in the textile domain but in the field of industry, the aeronautical and automobile industries in particular, and has actually provided one of the most solid materials in history. It is not possible to assign to it a precise date, nor a single inventor. The possibility of making polymerized fibres consisting of **long chains of carbon atoms** with very strong links was envisaged during the 1930s. The process consists in using heat to take away the appendices of ordinary **organic fibres**, for example the cellulose fibres like viscose rayon, so as to have only fibres of a specific molecular orientation, made entirely of carbon. This is carried out in precise temperature conditions by pyrolysis (decomposition by heat-ing); at one temperature, carbon fibres are obtained; at a high temperature, **graphite**, which is less resistant but a better conductor of electricity.

The carbon fibres, which are very stable mechanically and thermally, are used in the manufacture of **composite materials** which are meant to withstand significant forces, such as those of the compressor rotors of aviation turbines, the plating on spacecraft and the materials used for submarines which descend to the greatest depths. The first firm to produce carbon fibre was the British firm Courtaulds (in 1964), under the name of Graphil, after 20 years of research by the Atomic Energy Authority, Harwell. The American firm Hercules Inc. also produced carbon fibre in the same year.

Catalytic cracking of oil
Houdry, 1928

Cracking is the conversion, using heat, of **heavy hydrocarbons**, such as those which are directly extracted from the earth, into **light hydrocarbons**, such as those which are used as fuel. In chemistry, cracking is described in a more precise way as the splitting up of a saturated aliphatic hydrocarbon (also called a paraffin), into another paraffin and an olefin with a lower carbon number. In 1928 the Frenchman Eugène Houdry thought of achieving the splitting of primary paraffins by using a catalyst of **crystallized aluminosilicates**. These acidic substances cause the breakdown of carbon-carbon links. This system, which operates at temperatures of 500°C, gives a yield of 45–50% spirit with an octane number of between 91 and 93. The main problem is that this process causes coke to form at the surface of the catalyst, meaning that burning in air occurs.

This type of cracking is called **catalytic** and differs from **thermal**, which in principle is reserved for lighter hydrocarbons.

Cermets
Collective, from 1960

Cermets are **skeleton ceramic materials** into which metals are incorporated. They are new materials, capable of resisting very high temperatures due to their **carbide, borate** and **oxide** content. They are moulded together by **sintering** (baking) under very high pressures at temperatures lower than their melting point. Some of them are 70% aluminium and 30% chrome, others are based on chromium, titanium and zirconium borates, others on cobalt and titanium carbides. As well as their **refractive properties**, cermets have a high resistance to mechanical traction and they are chemically perfectly stable.

It is impossible to assign an original inventor to cermets, which are actually derivatives of ceramics in the ordinary sense of the word. It was from the 1960s onwards that metallic ceramics began to be used, the only materials stable enough to withstand the very high temperatures of rocket jets; indeed, no known alloy was suitable for the manufacture of the blades which guide the direction of these jets. They are also used in making the rotors for jet engine turbines and for coating spacecraft such as the US space shuttle, which has to withstand very high temperatures when it re-enters the Earth's atmosphere. The car industry expects the majority of cars to be made of cermets by the end of the 20th century.

Commercial oil pipeline
Van Syckel, 1863

The first commercial 'pipeline' seems to have been installed in Pennsylvania in 1863 by the American Samuel van Syckel. It ran between his oil field and a railway station six miles away.

Conductive plastics
Shrakawa, c.1975; McDiarmid and Heeger, 1977

The making of **electrically conductive plastics** resulted from an accidental discovery. It is a physico-chemical entity which would have seemed absurd in 1850 since one of the essential properties of plastics is precisely their non-conductivity or **insulating power**. In 1972 a researcher in the Hideki Shirakawa laboratory at the Tokyo Institute of Technology attempted to make a polymer, **polyacetylene**, which had been known since 1955 and which came in the form of a black powder. He obtained a completely different substance. He had made an error in adding a thousand times more **catalyst** than was required; he had made polyacetylene, but in a form which was previously unknown. The metallic look of the substance (like aluminium foil) and its electric conductivity were to intrigue Shirakawa, who began the study of what were then called **synthetic metals**. The next stage was reached in 1977 when Shirakawa and the Americans Alan G. McDiarmid and Alan J. Heeger, with whom he worked at the University of

Pennsylvania, had the idea of doping the new material with **iodine**. They obtained a kind of polyacetylene which this time resembled gold leaf and was a far better conductor. Thus began conductive plastics; ten years later there were a dozen of them. They consist of atoms of carbon and hydrogen which form into chains of **monomers** which in turn are shaped into polymers. The most conductive one is polyacetylene.

Contractile synthetic material
AIST, 1988

At the end of 1988 there was no generally accepted generic denomination for a material based on **polyvinyl alcohol gel**, which was made by the Agency of Industrial Science and Technology of Japan and which had the property of contracting in the presence of **acetone** and expanding in

the presence of **water**. Nor were there any precisely predicted uses for a material with such contraction properties. There is a possibility that this principle could be used to make **artificial limbs with muscular functions.**

Corrugated iron
Carpentier, 1853

The ribbing of sheet metal by a process of **stamping** was invented in 1853 by the Frenchman Pierre Carpentier. This material, called corrugated iron, was to

solve the problems incurred by **metal roofs**, particularly of warehouses and hangars, throughout the world. It has hardly changed since then.

Filisko fluid
Filisko, 1988

In an ordinary **hydraulic shock-absorber**, the fluid used has a given **viscosity** which limits the use of the whole system to determined temperatures. Indeed, the fluid cannot have the same properties at, for example, 0°C as it has at 100°C. In 1988 Frank Filisko, an American professor of metallurgical engineering from the University of Michigan, designed and made

a hydraulic fluid, the viscosity of which is altered according to the circumstances by a very weak electric current. An intense **electric field**, for example, causes the liquid to congeal, a weaker field halves its viscosity, etc. Thus the same fluid can withstand very different amounts of **pressure, tension** and **friction** by changing from a gel-like state to one of low viscosity in a very short space of time (up to 1 millisecond). Moreover, the alteration in viscosity can be controlled by a micro-computer.

Since it is non-toxic and capable of withstanding temperatures in the region of 150°C, the Filisko fluid is convenient to prepare since all that has to be done is to pour a powder, the nature of which is protected by a patent, into any liquid of satisfactory original viscosity, such as mineral oil. An American 'quart' (about one litre), was commericialized in 1988 at a price of $3, about £1.80. Filisko fluid, which is the property of the Michigan firm Tremec Trading Co., could cause considerable changes in all hydraulic systems, notably in the construction of motor cars.

Fluid jet cutting
Yi Hoh Pao, 1970

The first effective cutting tool which used a jet of fluid under pressure was invented in 1970 by the American Yi Hoh Pao, a former employee of the Boeing company. It consists of a nozzle which ejects a liquid at 800 m/s under a pressure of up to 4000 bars, and it can cut a block of concrete with good accuracy. The idea of using a jet of water for cutting a solid object dates back some time, for at the beginning of the century at the time of the Gold Rush, some engineers caused hill erosion using pressurized water jets. Around 1930 hydraulic pressure amplifiers existed on the market and could have enabled the manufacture of hydraulic cutting nozzles; the difficulty which opposed this was the **vaporization of the water** which was coming out at such high pressures. Credit is due to Yi Hoh Pao for incorporating a **long chain polymer** into the water which guarded against the vaporization.

Grafted textiles
US Army, C.1950; ITF, c.1963

Grafted textiles are essentially textiles made of **polymers** which are modified according to specific researched properties, although the 'graft' can take place on natural fibres. In principle the modification of the molecule consists of breaking the chain of molecules which make up the polymer in order to insert between the broken bonds an **unstable monomer**, the 'transplant', the fixing of which is facilitated by a double unstable bond. This technique therefore confers some extremely varied properties to the textile, which are described below.

It is difficult to assign a date and an author to the origin of this invention, which seems to have taken place in the United States during the 1950s but was thenceforth covered by military secrecy; its beneficiary is in fact the US Army. As far as one knows, the technique was aimed at the manufacture of **crease-resistant clothes** for military staff, a technique which since then has largely taken over civilian clothing. It would seem that the research begun in America was not actively followed up there. In 1963, however, the ITF, or Textile Institute of Lyon, took up the research for itself and made considerable progress.

The research was first of all aimed at clothing textiles. It resulted in processes which assured fast dyes and fibres with a particular 'look' or a specific texture, such as being silky or woolly. It also notably included **hydrophobic fibres** which naturally repel water, and **fireproof fibres** (numerous national and foreign regulations demand that certain materials, children's pyjamas, seat covers in public places, aeroplanes and cars etc., be treated against fire).

The ITF uses four different grafting processes: **chemical activation**, which is preferable for natural fibres; **ionizing radiation** using an electron gun; **ozonisation**, which in 1988 was still in the experimental stage due to problems regarding the use of ozone; and using **cold plasma**. This latter process differs from the others by the fact that it is not the polymers which are activated but the monomer, which can then break the polymer to insert itself.

This research, which has very appreciably lowered the manufacturing costs of common textiles, has spread to areas beyond the ordinary uses of textiles. It was in this way that **purifying filters** were thought of, which could trap specific pollutants, such as fuel oil, chemical products, toxic gases, etc. It is also possible that a new type of catalytic exhaust will appear in the car industry at the beginning of the 1990s.

In the medical domain the ITF in collaboration with the Pasteur Institute is expected to produce some **autosterile, bactericidal** or **haemostatic textiles**. In the USSR parallel research since 1985 has enabled the manufacture of haemostatic gauze dressings (to stop bleeding). Finally, insulating materials can be imagined, or alternatively electrical conductors or, in the fashion world, materials which change colour according to temperature. Grafted textiles, which represent the 'second generation' of synthetic textiles are discussed under **cryptans** (see p. 27).

Heat-resistant glass
Zeiss, 1884

It was in 1884 that the German Carl Zeiss made a **borosilicate glass**, which was particularly rich in **boric acid** and so had exceptional qualities of **heat resistance**.

These results led to the manufacture of particular types of heat-resistant glass such as **pyrex**.

Industrial fibreglass
Owens Illinois Glass Company, 1931; Chrevolet, 1953

Although it was invented in the 18th century (see *Great Inventions Through History*), for a long time fibreglass was thought of as a curiosity without much of a future. So we can refer to its re-invention when it was industrially produced in 1931 by the American firm Owens Illinois Glass Company. Fibreglass material began to be used on a large scale for its heat-insulation properties. It is difficult to identify precisely the engineer who verified its mechanical properties, but the first to use fibreglass to make the entire bodywork of cars was the American firm Chevrolet in 1953. The pliability of the new material enabled the firm to make, at a reduced cost, frames whose striking profiles would have demanded much more expensive work to be done on traditional metal panels; they were also incomparably light in weight. The first model built in this way gave rise to the Corvette series of sports cars. The construction of fibreglass **streamlining** quickly won over the **pleasure boats** industry.

Kevlar
Kwolek and Du Pont de Nemours, 1964–73

Kevlar is an exceptionally resistant fibre of the **polyamide group** and is made of polyparaphenyleneterephthalamide. It was created by the American Stephanie Kwolek and the researchers from the American firm Du Pont de Nemours. The federal offices were provided with it in 1969 under the name of PRD-49 and it was commercialized on a larger scale in 1973.

Laser scissor for textiles
Hughes Aircraft, 1980

The laser scissor, which is used largely in the **clothing industry**, is capable of cutting cleanly through several layers of material. It was made in 1980 by the American firm Hughes Aircraft Corp.

Latex foam
Murphy, Chapman and Dunlop, 1929

In order to make **rubber**, which is quite heavy in its compact form, lighter, E. A. Murphy, an American engineer from the Dunlop Latex Laboratories, had the idea of beating the latex into a foam with a kitchen whisk, as if it were egg-white. He did obtain a latex foam, which was lighter than compact rubber because of the air bubbles, but it did not 'hold' the air until it hardened. It was his colleague W. H. Chapman who discovered the process which enabled the foam to be polymerized at the point when it contained the maximum number of air bubbles, and, more significantly, it was he who had the idea of pouring this mousse into moulds. This was the starting point for Dunlopillo, which was to cause a revolution in industrial upholstery (car seats and auditoriums) and in the craft industry.

Memory alloys
Buehler, 1962

Memory alloys were discovered during the 1930s. Their name comes from the fact that they are shaped and formed under heat, and although they lose their original form when they cool, they regain it when they are reheated. Their transformation can be explained by the modification which is undergone, according to the temperature, by their **crystalline structure**. That is, as it passes from the **austenitic phase**, stable at high temperature, to the **martensitic phase**, stable at low temperature, depending on the metal remaining in a solid state from one phase to the other. In 1962 the American W. Buehler, from the Naval Ordnance Laboratory, discovered that this feature was present in an alloy of equal parts of titanium and nickel, called **Nitinol**, and used it in some experimental structures. Its first 'official' use was by the National Aeronautics & Space Administration (NASA) in the making of a **satellite aerial** which was rolled up in a ball when launched and straightened out as soon as an electric current was passed through it.

Memory alloys are characterized by their **elasticity**, which is up to ten times greater than that of ordinary metals and alloys. They present the possibility of electronic components allowing emission of different messages depending on their temperature. There are also ideas for technological equipment such as taps which turn themselves off when the water becomes too hot. In 1988, Pechiney made alloys which had double memories, that is, they took different shapes at three distinct temperature levels.

Memory plastics

Nippon Zeon, 1987

Memory plastics are synthetic materials which change their shape in a particular way according to the temperature. Once deformed at a given temperature, they will regain their original shape at another temperature and keep it. The first type was an **elastomer** which was discovered in 1987 by the Japanese firm Nippon Zeon. When heated to a temperature of 37°C this material takes on that which could be called its 'mother shape', then it is cooled. The various deformations, tears and alterations which it then undergoes are obliterated as soon as the material once again reaches 37°C, the temperature at which its integrity is guaranteed. The phenomenon can be explained by a large **molecular density**. Above 37°C the memory behaves like a rubber; below, it shrinks and hardens. By the end of the 1980s the exploration of the possible uses was still in progress. The inventors were thinking, for example, of making studded tyres; the studs would be invisible at temperatures over 0°C, but the shrinking of the material would make them appear at this temperature. Other uses also in the car industry

have been envisaged, particularly concerning the body of the car. The parts of the body most likely to receive knocks could regain their shape after simply being heated to the original temperature.

In order to understand that some plastics are conductive, as well as the significance of this phenomenon, it must be appreciated that this property is linked to their atomic structure. The electrons are able to diffuse only between the atomic bonds in a structure. In doing this they cross over energy levels, called **bands**, which have determined capacities for absorbing electrons. They can only cross bands of specific energy levels, that is, those which are neither saturated nor empty. Metal conductors are characterized by bands which are favourable to the exchange of electrons; other materials are non-conductive because their bands are either empty or saturated. The bands in conductive plastics are suitable for the conduction of electricity because of their atomic structure which differs from that of ordinary plastics, a difference which is due to the dopant (iodine in the case of polyacetylene).

At the end of the 1980s a complete theory of the conductivity in conductive plastics was still being researched; it was not yet established. One of the factors of the explanation might even possibly be the polymer

The damage caused by a revolver bullet in a memory plastic disappeared when heated so that it was invisible to the naked eye.

chains. Once the theory has been established it is thought that **superconductors, photoelectric cells** and **solar batteries** will be developed from these plastics, which have the property of light absorption; in the medical domain **artificial limbs** (of

polypyrrole) are expected.

As they are equally easily woven as dissolved in liquids, the conductive plastics have opened up many scientific pathways which during the 1980s could only be partially understood.

Non-greasy lubricant
Du Pont, 1961

Specializing in the production of synthetic materials, the American firm Du Pont produced a **fluorine composite** which was non-greasy and heat resistant, and which adhered to its supporting medium but also impregnated any surface in contact with it.

Being non-greasy, this material had the advantage of not becoming clogged up with dust during use. It was the first of the non-greasy solid lubricants and it paved the way for the invention of Teflon.

Optical fibres
Kapany, c.1955

Optical fibres (the term is used in the plural) is the name for transparent **glass-fibres** which contain a **core** of higher refractive index than the **cladding**. Rays of light travel through them by successive reflections. The invention is derived from glassfibres, the first of which were made in the 19th century.

It was the Indian Narinder S. Kapany, later an American citizen, who in 1955 thought of using glassfibres encased in a cladding to direct light along a non-rectilinear path. This principle was first of all used in the manufacture of **endoscopes** for medical exploration. Being **optical wave guides**, these fibres were also the object of studies which proposed to use them for long distance communication. The principle for this is as follows: at

transmission a **modulator** transforms the electric signals into light signals and at reception these are transformed by photo-diodes into electrical signals. The communication following the signals can be modulated light impulses (digital) or analogue wave forms. One technical problem existed: in order to travel long distances the light signal had to use a very pure medium, a fibre which would absorb very little light. Such a fibre did not exist. However it became possible in 1960 when the American Corning Glass Works found a way of making very pure glass. From the mid-1970s onwards the transmission of **video signals** by optical fibres was an established fact. Optical fibres are the main medium for carrying telecommunications of all types.

Optical sound recording

De Forest, 1920; Philips, 1979; 3 M Inc., 1981

The first inventor to think of transforming sound waves into light impulses of corresponding intensity using a **photoelectric cell** (a device to convert light into electricity) on a sensitive film, and of then using the inverse process to transcribe the images obtained into sound, was the American Lee De Forest in 1920. He undoubtedly had a predecessor in this field, the Englishman William Du Bois Duddell, who in 1901 had made a sound recording on a sensitive surface. The De Forest process was the first, however, which could be used for the transcription of sounds which were synchronized with images on a lateral track for the cinematography industry of the importance of his invention, but in the first place it was not complete, and also, De Forest had no business sense and was duped once again: a few years later a process strangely similar to his own appeared and 'talkies' were launched.

For a long time optical sound recording was used only in the cinema, but at the beginning of the 1970s numerous inventors endeavoured to apply it to music records. In 1979 Philips launched the **compact disc** which has optical reading. It differed from the De Forest process in that the sound was no longer recorded in an **analogical** way.

Instead, it was coded in a **numerical** form and the micro-hollows on the disc were processed at a rate of four million per second; this process, which due to the high cost of the players and the discs themselves took several years to become widely used, produced a much better clarity of sound than the microgroove.

In 1981 the American firm 3 M Inc., from St Paul (Minnesota), developed a reliable process for the production of **alloys of rare ores** which could withstand the laser engraving temperature of just 150°C without alteration of the chemical structure, that is, with neither oxidation nor decomposition. This enabled the same disc to be used for new recordings several million times over. The **erasable optical disc** was born. Moreover, these alloys, gadolinium–terbium–iron, terbium–iron–cobalt or terbium–iron, have a quality of sound reproduction superior to previous means of sound recording.

> The erasable optical discs are called **magneto-optical** because the magnetism plays an important part in their use: in order to erase the data all that is required is for the magnetic field to be reversed (North downwards).

Pneumatic drill

Sommeiller, 1861

The idea of using **compressed air** to release bursts of impact was conceived by the French engineer, Germain Sommeiller.

This new kind of tool was to prove itself invaluable during drilling of the tunnel through Mont Blanc.

Ranque tube

Ranque, 1928

Having studied **turbulent flow** since 1922, in 1928 the Frenchman Georges Ranque made a tube with a vortex and a needle, which was later to be developed as the tube named after him. **Compressed air** is fed in to the tube and since its pressure decreases the tube is a cooling centre. When the hot air hits a needle at the opposite end to the direction of injection, it separates into two flows. One of these, which is hot, is ejected, whereas the other, which is reflected, gradually cools down. Though used in industry, the Ranque tube remained misunderstood for a long time. Then it was rediscovered in 1945 by the German Rudolf Hilsch, who did refer to Ranque's work but gave his own name to the tube; so it is sometimes known under the name of **Hilsch tube** and sometimes even under that of **Vortex tube** (see diagram below).

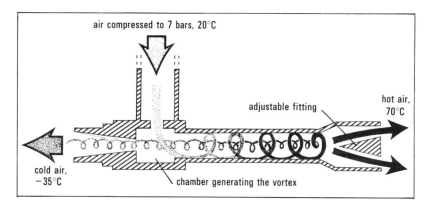

The Ranque tube is based on a disarmingly simple principle: if compressed air is injected into a tube like the one above, it causes a clockwise spiral flow. As it moves forward this flow heats up and at the exist it separates into two flows, one which leaves the tube and another which goes bac again, still spiralling but within the first spiral flow. As it makes its way in the opposite direction this second swirl cools down; it exists, on the left, at various temperatures between −35 and −75°C.

Reinforced concrete

Hennebique, 1892

Contrary to common opinion, concrete is very old, for it was invented by the Romans (see *Great Inventions Through History*). The 'secret' of it was lost during the Middle Ages, and was not found again during the Renaissance, but it was gradually rediscovered in the 18th century, firstly by the Englishman John Smeaton who made a kind of 'Roman cement' based on calcined clay shingle, then by the Germans Lippman and Schneckenburger who were the first in 1859 to attempt to make artificial marble.

The invention of reinforced cement in 1848 was to lead to the invention of reinforced concrete by the Frenchman Hennebique in 1892. It consists of hydraulic concrete into which a **metallic framework** has been inserted during casting so that blocks are obtained which have a **bending** and **tensile** strength far superior to that of ordinary concrete. Hennebique thought that reinforced concrete would therefore offer better protection from fire. Experience has shown that the internal expansion of the metal framework under the effect of heat can actually cause bending and traction which make the concrete around it explode. In this particular respect, plaster still constitutes one of the best among the classical materials for delaying the effects of fire.

Solar paint

Los Alamos National Laboratory, 1985

A type of paint capable of capturing **solar energy** was made in 1985 in the Los Alamos National Laboratory of the United States. It contains metallic particles which slow down the loss of heat. Compared to the traditional black paints it is 10–20% more productive.

Stainless steel

Brearley, 1913

So-called stainless steel is not absolutely resistant to oxidation as its name would suggest, but it does offer sufficient resistance to alkalis and ordinary acids to partially justify its name. It was both discovered and made in 1913 in Sheffield by the Englishman Henry Brearley, using an **alloy of steel and chrome**, and it was immediately a great success. The following year, the firm Krupp made a steel which was even more resistant and consisted of a steel alloy of 18% **chrome** and 8% **nickel**. The resistance to corrosion depends on the chrome content. The **martensitic** steels are obtained through shearing and contain 12–18% chrome and 0.12–1% carbon; the **austenitic** steels, which are based on **gamma-iron** and **carbon**, are made according to the Krupp alloy formula mentioned above.

Stamp of origin

Anon, UK, 1887

In 1887 the British Parliament were concerned about the strong competition from German industry, represented by Alfred Krupp and Werner von Siemens, and passed the **Merchandise Marks Act** which stated that any product manufactured in Germany and crossing the borders of the British Empire should carry the label *Made in Germany*. Subsequently such a label was required of all foreign merchandise. Parliament hoped that this would encourage British subjects to have a preference for national products, a hope which was to be unrealized.

Stapler
Gould, 1868

The first automatic stapling system was invented in 1868 for bookbinders by the Englishman Charles Henry Gould. By saving them the need to sew the periodicals together, the stapler increased the delivery speed of numerous publications.

Stereolithography
3 D Systems, 1987

The American firm 3 D Systems Inc. invented and patented this system in 1987 and gave it the misnomer; it is really to do with rapid **stereoplasticity**. The process consists of making plastic **prototypes** of mechanical things. This takes from a few minutes to several hours, a considerably shorter time than is usually required for these prototypes and is achieved by using a computer to draw the prototype from all sides, following the required parameters. These three-dimensional parameters are then followed by a moving **ultra-violet laser** which is guided over a reservoir of transparent liquid plastic; the plastic is instantly polymerized and all that remains is for it to be taken out of the liquid, or else, if it is only a part of a prototype, for the different elements to be assembled using the same technique. In this latter case, the whole thing is covered with a layer of polymerized plastic.

When it was first commercialized, this system, called SLA/I, could only treat prototypes with a maximum surface area of 30 cm^2 and its precision was limited due to the shrinking of the plastic.

Toughened glass
Appert, 1893

The expansion of iron architecture where many windows were used inspired the Frenchman Léon Appert to invent toughened glass. It has an internal metallic structure following a principle similar to that of reinforced concrete which was discovered the previous year.

Tungsten carbide (industrial production of)

Moissan, 1907

It was in 1893 that the French chemist Henri Moissan had the idea of using the **electric oven** to achieve the fusion of numerous metallic oxides such as chrome and titanium, as well as carbides, hydrides, nitrites, silicates and crystallized borates. This work was to go on for several years and initiated the production of tungsten carbide around 1907. This inorganic composite of carbon is essential to the fabrication of **hard steels**, drills, saws and high-resistance plating; it is made by heating tungsten powder with an ash to temperatures between 1400 and 1600°C. In 1927, the German firm Krupp and the American firm General Motors applied and developed the Moissan process to the production of **cobalt tungsten carbides**, now known under their registered trademarks Widia (Krupp) and Carboloy (GM).

Ventilator

Guibal, 1858; Ser, 1878

The principle behind the aeration ventilator seems to derive quite naturally from the windmill, which is very old, and from turbines, which date back at least to the 17th century. But in fact one of the peculiarities of the history of technology is that in the mid-19th century when the ventilating of public places, auditoriums and worksites was a problem, not even a simple apparatus such as one made of **paddles on a shaft**, driven by steam, for example, had been devised, either for clearing stale air or drawing in fresh air.

Mines were among the places where

It is interesting to note that the blade ventilator which took such a long time to be adopted, had been used in 1553 for the aeration of mines by the German Georg Bauer, who is better known under the Latin name of Georgius Agricola. It was a hand-operated machine. For centuries the reasons why it was necessary to aerate the mines was discussed. Some argued that it was because of the heat, others that it was because of the restricted space. In 1777 Lavoisier suggested that it was due to the build-up of **carbon dioxide** in the occupied spaces, but the German Max Joseph von Pettenkofer proved in 1863 that this represented too small a fraction of the volume of air, 0.03–0.05%, to explain the observed results and that the corresponding decrease of **oxygen** was 1% and thus also insufficient to explain the need for aeration. Around 1885 the French doctor Charles Edmond

Brown-Séquard put forward the theory, which now seems rather strange, according to which the noxious effects of confined air were due to that which he called **anthropotoxins** or **morbific matter**, which was not without reference to the **miasmas** which, until Pasteur, his colleagues had blamed for the spreading of contagious diseases. It was not until the beginning of the 20th century that the German Karl Flügge, who is still remembered in connection with the droplets which were named after him, proved that the harmful effects of the lack of aeration come from the inhibition of the natural heat-loss through radiation of heat from the human body. (**Flügge's droplets** are those which consist of saliva and are usually vaporized in the air during speech, carrying germs capable of contaminating the person being spoken to.)

a ventilation system was most urgently needed. Even in the shallow ones the air quickly became unbreathable and in the 18th century **pumps** were designed and made which injected fresh air up to a distance of about 500 m. Obviously this was better than nothing but as a result of their inadequate effect — the injection of fresh air was hardly even enough to de-pollute the air — a replacement was sought; for some time this was found in the lighting of **fires** at the shaft exits, where they created in-draughts. However it did not hold the solution to **firedamp** (a combustible gas found in mines) which was responsible for innumerable and fatal explosions. In 1838 the Frenchman Charles Combs had published the first scientific analysis of the problem, where it was pointed out that first of all the stale air must be taken out of the mines, and that air must be made to circulate from below in an upward direction by specially drilled **airtight chimneys**.

Since 1820 pumps had been used to aerate mines which compressed the air inside the mines to make expelling it easier. Their productivity left much to be desired and, apart from the inconvenience of their bulk, which was considerable, the ventilators could cause great inconvenience by blocking the aeration if they stopped.

The solution came at last in 1858 when the Frenchman Théophile Guibal invented the **dynamic ventilator**, called the 'encased paddles', which captured the air in the mines through a chimney inside an airtight chamber and then channelled it to the outside. It was immediately adopted for use in mines throughout the whole world.

Twenty years later the Frenchman Louis Ser published the first **analytic theory of ventilators**, which pertained to the mechanics of fluids.

Instruments for Measuring
and Observation

A record of all the measuring and observation instruments invented since 1850 would need a book of its own, even if only a few lines were allotted to each.

The inventions presented in this chapter are therefore considered to be the most outstanding. Without the interferometer, of which there are about 20 variations all with specific purposes, our knowledge of the cosmos would be much poorer and theoretical physics would suffer severely. Without the camera lens, photography would not exist. Without radiography, medicine too would be in difficulties. And these are only a few examples.

Progress in this domain began in the 17th century; these instruments seem to answer a human need to perceive that which lies beyond the potential of our senses. There is a desire first to satisfy the irrepressible instinct for discovery and then, to verify the measurements which have been made of the physical world since early history. Scientists, like other people, have never been satisfied with the Platonic parable which proposes that what we humans see of the world is only a projection of shadows at the base of a cavern.

The measuring and observation instruments constitute the milestones of an adventure. The Greeks began by evaluating the circumference of the Earth, the French verified and clarified it, and then established the standard metre, and nowadays the ångström is the unit of measurement.

Adding (or calculating) machine

Kelvin, 1879; Hollerith, 1880

In the mid-19th century the masterful ideas concerning the adding machine, later to be called a **'computer'**, could be traced to several men: Pascal, who devised the mechanical infrastructure; Leibniz, who extended the capacity of Pascal's machine which was limited in its counting ability to the level of the arithmetical operations of multiplication, division and square roots; Vaucanson, who developed the programs on perforated or punched cards; Babbage, who invented the analytical machine; and George Boole. In 1847 Boole had published a major and precursory work, *Mathematical analysis of logic*. In this he studied the fundamental laws of the intellectual processes of reasoning, expressing them in mathematical language and laying down the basis of logical science. This was a significant work because it both integrated logical relationships fully, and enabled logic to be introduced into the adding machines.

Until then logic was based on the use of **syllogism** and **deduction** and was thus limited to the domain of philosophy, but it was absorbed by mathematics and therein found a new unsuspected strength. Just as Boole had said, since reasoning could be assimilated to an algebraic form of calculation and mathematical rules could be applied to solve logical problems, it became possible to enter the rules of logic into these adding machines and thereby considerably extend their range of ability. Nevertheless it must be said here that Boole, a logician, did not actually assimilate logic to mathematics: he put it in parallel and expressed its laws in mathematical form, as a way of explaining the logical figures and deductive reasoning. In doing this, Boole created the algebra which is now named after him and invented **binary language**. Thenceforth the algebraic operations of logic could take place using only two numbers, 0 and 1. Boole's teaching fascinated the mathematical and philosophical world for half a century. In the same way as Babbage, Boole tried to build a machine which would incorporate these principles but he did not succeed.

In 1879 Boole's project was taken up again by Sir William Thomson (Lord Kelvin), who managed to bypass the difficulty inherent in the making of a digital machine by making an analogue machine. Kelvin's device was meant to solve **finite equations**; it consists of eight pulleys which are moved by adjustable handles — four pulleys on a level above a wooden frame, and four on the level below. Two pulleys, one above and one below, are set in motion by a cord attached to a weight and a marker; given that each pulley follows a circular movement of adjustable size, which is equivalent to the sum of two simple or sinusoid harmonics, one horizontal and one vertical, the operation takes place in the following way: the horizontal component of the circular movement tends to make the cord leave its vertical position, but if the range of movement followed by each pulley is only a fraction of the distance between the two pulleys, then the horizontal component has a weak effect; the principal action on the cord is that of a vertical

The history of the calculation machine comprises many more names and experiments than could ever be included here. It is appropriate nevertheless to mention the Englishman Williams Jevons, an admirer of Boole, who built a **'logical piano'** based on Boole's principles. He might have made a considerable contribution to the genesis of computers if he had not drowned in 1882 at the age of 47. The torch was carried by Lord Kelvin.

The case of George Boole has often served as an example in the argument for inherited intelligence. Indeed, from the logician's marriage with Mary Everest, the niece of the geographer in whose honour the famous Himalayan mountain was named, five girls were born. Of these, Alicia became a mathematician and Lucy was the first woman chemistry professor in England.

sinusoidal component. The weight hanging at the end of the cord will therefore follow a movement which is the **sum of the vertical components** of the two pulleys.

Kelvin's machine was in fact not a calculator but a **mathematical machine** which could, for example, forecast the tidal movements for a year. It was also the first **analogue computer**. A reliable version of it was not made until 1930 at the Massachusetts Institute of Technology.

In 1880, thanks to the American Hermann Hollerith, the digital adding machine made its official entry into government services. Hollerith, a specialist of statistical studies,

improved the principle of **punched cards** and made a machine which allowed population censuses to be recorded five times more quickly than before. Indeed, each card could carry 16 five-digit numbers or 8 ten-digit numbers. It provided the starting point for the computer IBM. The difference between the adding machine and the computer must be established (see p. 73). Relatively speaking, it can be said that the adding machine set out the basis of logic, the use of which the computer was going to implement, and that inside every computer there is first of all a calculating machine.

Anthropometry
Quételet, 1871; Bertillon, 1880

Anthropometry is the art of measuring the different parts of the human body and of then eventually defining a **type** from their proportions. It is said that the origins of anthropometry date back to the **aesthetic Greek canons** which first assigned a numerical value to the size of the head in proportion to the rest of the body. However it seems that the first to have thought of anthropometry as a separate discipline was the Belgian mathematician, astronomer and statistician, Adolphe Quételet, in his work *L'Anthropométrie, ou mesure des différentes facultés de l'homme* which was published in 1871. Quételet, with whose name his colleague André Michel Guerry should be associated, intended to apply the laws of **statistics** to the whole set of biological, intellectual, and social phenomena of humanity and the physical and intellectual growth which comprise mortality.

This idea was to be taken up again in criminal anthropology by the Italian Cesare Lombroso from the point of view of **phrenology**, a discipline which postulates that certain mental characteristics, particularly the propensity to crime, correspond to specific cranial shapes. Phrenology was founded at the beginning of the 19th century by the German Franz Joseph Gall.

This offshoot of anthropometry must be

mentioned here, despite its having little scientific basis, because it has inspired theories and research which are all more or less destined to prove Gall's unfounded ideas, particularly the existence of **a relation between cranial capacity and intelligence**. Despite formal refutations brought against this use of anthropometry, such as those of the American Kasmin, and despite also the known distortions of which one of the most famous defenders of the theory, the Englishman Sir Cyril Burt, was guilty, this non-scientific derivation still holds water in certain circles.

Anthropometry makes a study of the main dimensions of the human body, particularly weight, height, skinfold thickness, mid-arm circumference and waist-to-hip ratio. One of the first uses was for the selection of recruits for military service.

Around 1880 the Frenchman Alphonse Bertillon developed anthropometry and one aspect in particular, **cephalometry** (measurement of the skull), in order to found **judiciary anthropometry** which was aimed at identifying criminals. Since he was Head of the Identification Service at the Paris Police Headquarters, Bertillon thought of using anthropometric data to recognize criminals, and also of dividing the data into three groups (called trichotomic), with all three arranged according to the length of the head, which seemed the most reliable measurement. Then he created three other groups based on the width of the head, with each sub-group being characterized by marks such as the length of the left middle finger, then the left little finger, etc. This is called the 'Bertillonage' method of classification. Purkinje's discovery of fingerprints was to improve the methods of anthropometry considerably.

This invention has been unquestionably useful in several disciplines. In **anthropology** it has enabled the definition of the characteristics of the ethnic groups of the human race. In the first half of the 20th century it had been improved with numerous criteria, such as the thickness of adipose (fat) tissue, blood capillaries and the blood groups.

In **medicine**, anthropometry has allowed correlations between certain body types and certain diseases to be established, notably those illnesses concerning nutrition and the cardiovascular system, correlations which have caused the drafting of assurance policies to be altered according to anthropometric information.

During the 1980s the refining of anthropometric methods associated with the use of a **synthetic image** treated by a computer gave rise to a numerical technique of reconstructing the appearance of a missing human being according to traces such as the skull or the femur. This new technique has benefited both archeological anthropometry and judiciary anthropometry at once (the latter for the identification of recent skeletons).

Atomic-force microscope (AFM)
Binnig and Rohrer, IBM, 1985

The atomic force microscope is derived from the **scanning tunnelling microscope** (see p. 157) and is based on the same principle, but on such a small scale that it is no longer the electric current but the electrons themselves which enable the exploration of the surface in question due to the **force of atomic interaction**. The AFM used a splinter of diamond to explore the surface of the material being examined.

The needle traced the surface of the material — the tracking force is so small that the diamond tip can trace individual atoms without damaging the surface. This microscope was invented by G. Binnig and H. Rohrer in the IBM laboratory in Zurich and has the advantage over the STM of being able to uncover structure of **non-conducting surfaces**, so it can be used for imaging biological molecules.

Audiometer
Hughes, 1879

An audiometer is an instrument which serves to measure the acuity of **human hearing**. In principle it consists of an oscillator, the amplified output of which is communicated to earphones or else to a vibrator (for measuring the ability of the ear bones to transmit sound). The frequency and intensity of the sounds can be adjusted,

since the measurement of auditive acuity is made according to the emitted sounds. The first of these instruments was invented in 1879 by David E. Hughes; it consisted simply of a set of two 'Leclanché cells' (batteries) connected to a microphone and a telephone.

Cash register

Burroughs, 1892

The cash register for commercial use, with a **display screen** was developed from Hollerith's **calculating machine** (see p. 139). It was invented in 1892 by the American William Seward Burroughs, founder of the American Arithmometer Company which was succeeded in 1905 by the Burroughs Adding Machine Company. Burroughs never saw the success of his invention, for he died in 1898.

Chronobiology

Brüning and Pittendrigh, c.1945

'Chronobiology' is the quantitative study of the **rhythmic changes of biological phenomena**. It plays a growing role in medicine, notably in the choice of the administration time of medication, given that the effect of the same dose can vary considerably according to the time of day at which it is taken.

Chronobiology is derived from the already ancient observation of the rhythms which control the functioning processes of living things, whether vegetable or animal. It is thought that it originated with the timetable of the opening and closing of the flowers of various plant species which was drawn up in the 18th century by the Swedish naturalist Carl Linnaeus. This work was taken up again by the Frenchman H. Duhamel du Monceau who grew flowers at a constant temperature in the dark. Since then many observations have been made in the 'biological' studies carried out by naturalists, concerning the daily (circadian) or seasonal rhythmic functioning of animal and plant species; for instance, what makes the courtship season of foxes begin and end on a fixed date.

Around 1945 the German Erwin Brüning and the American Colin Stephenson Pittendrigh independently envisaged the existence of one specific chronological structure for a species, which they based on **genetic components**.

However it was following the work of the Frenchmen Alain Reinberg and F. Halberg that chronobiology really became the discipline of the interpretation of biological rhythms which was applicable to **pathology** (study of disease) in medicine. It is known today that the cardiovascular (heart and blood vessels), endocrine (secretory glands), digestive and nervous systems etc., undergo regular cycles of which the majority are **circadian**, and some are **ultradian** (more than two and a half days) and other are **infradian** (less than two hours); any change in these routines incurs pathological phenomena (i.e. illness). The research of the speleologist (researcher into flora and fauna of caves) Michel Siffre carried out at the bottom of a chasm confirmed the hypothesized existence of an **internal (body) clock**, the rhythm of which tends to be about an hour and a half faster than the astronomical clock.

During the 1960s two particular disciplines developed: **chronopharmacology** which, 20 years later, has led to the fixing of precise administration times for medicine,

Numerous theories proposing tables of psychological rhythms have been put forward since the 1920s. Sigmund Freud's doctor, Wilhelm Fliess, presented several models to his illustrious client who spent some time using them as a basis for trying to calculate the day of his death. The idea of rhythms of the nervous system has led to the establishment of periodic tables of the temperament. In several countries the **suicide rate** is highest in the spring and lowest in winter.

in order to achieve the maximum effect of minimum doses, and **chronotoxicology** by which the effects of natural and artificial toxins can be ameliorated, depending on the time.

The practical consequences are immense, influencing, for example, the organization charts of airline crews. At the end of the 1980s, chronobiology was also leading to the revision of certain job timetables such as atomic power station controllers and railway drivers, with regard to the existence of rhythms of intellectual vigilance.

Cloud chamber

Anderson and Millikan, 1930

The **cloud chamber** is used for the detection of **gamma rays** (high-energy radiation) coming from outside the Earth's atmosphere. It was invented in 1930 by the American physicists Carl David Anderson and Robert Millikan and consists of a gas-filled chamber containing a vapour at saturation point. The vapour is suddenly cooled by expansion, and condenses onto the ionized particles produced by the passage of a gamma ray through the chamber.

This appears on the photographic plate in the shape of a trail of liquid drops corresponding to the path taken by the particle.

The **bubble chamber** is a similar device, consisting of liquid hydrogen under pressure at a temperature slightly above its normal boiling point. If the pressure is released, bubbles form on the charged particles left by the passage of the gamma ray.

Dow-Jones index

Dow, Jones 1882

The day-to-day **index of Stock Market values** was inspired by statistical calculations and invented in 1882 by the Americans Charles Henry Dow, founder of the *Wall Street Journal*, and Edward D. Jones. It enables the rise and fall of the New York Stock Market to be estimated. There are equivalent systems in stock exchanges all over the world.

Electron microscopy

Abbe, 1873

Although developed during the first third of the 20th century, electron microscopy was started by an observation by the German Ernst Abbe in 1873. This observation contained the germ of electron microscopy as it showed for the first time that whatever the improvements made in optical microscopy, the fact could never change that two separate objects cannot be distinguished if the distance between them is less than half of the wavelength of the light which illuminates them. This is **Abbe's law**. X-rays, which were discovered in 1892 and which have a much smaller wavelength than visible light, ought to have provided the means of going beyond the limits of optical microscopy. However optics capable of focusing X-rays did not yet exist.

Futurology

Berger, 1925–60

Futurology is a controversial area of study which aims to discern the shape of future developments (technical, scientific, economic and social) on the basis of analysing the present and the past. It was invented by the French philosopher Gaston Berger who first thought of it around 1925 and never stopped refining its parameters. Futurology is sometimes considered to be a science although it does not correspond to one of the conditions which is essential to any science, the possibility of experimentation.

Futurology has been appreciably useful since the middle of the 20th century, notably in the economic and military domains. When applied to **statistics** it now constitutes a whole separate discipline in **budget development and the procurement politics** of all national administrations. Although it is not scientifically precise, futurology allows the forecasting of both expansion and recession in certain sectors, and so politics can be adjusted to the needs of a nation. It is also used by all the large industrial, economic and financial firms for this purpose.

Futurology has also given rise to **military scenarios**, which are used mainly by the military strategists of numerous nations to try to determine the outcome of a conflict according to the elements and incidents concerned.

Futurology at the end of the 1980s showed itself to be an approximate means of predicting the development of societies, but not an exact one. Two major events in particular have indicated the inadequacies of futurology. The first was the 1973 **oil crisis**, unforeseen by all the futurologists in the Western world, which suddenly found itself confronted with a crisis situation which it did not have the immediate means to resolve. The second was the abundance of oil in world markets which thwarted the plans for intensive procurement of atomic power stations being made by countries such as France which have no oil resources. This also held the price of the oil kilowatt-hour far below the scale of charges expected for a kilowatt-hour generated by a nuclear power station.

Geiger counter
Geiger and Müller, 1928

A Geiger, or more specifically, Geiger–Müller radiation counter is an **ionization chamber** which is for measuring **alpha** or **beta radiation** (which consist of charged particles). The principle behind it is simple: it is a cylindrical argon-filled chamber in which a very fine metal wire is stretched tightly along the cylinder axis. This wire is brought to a high electric potential. When a particle passes, the intense electric field near the wire incites the particle to ionize the atoms of gas; this is followed by an avalanche of ions which induces an electric current in the wire, which is amplified and measured by electronic circuits. The level of radioactivity can be read on a graduated dial. The idea behind this dates back to 1908 but it was 20 years later that Hans Geiger and Erwin Müller, professors of physics at Tübingen University, made a counter of this type.

In 1958 the American G. Charpak extended the principle of the one-wire ionization chamber to make a **multi-wire chamber** which measured 5 m in diameter.

Holographic computer
Anderson, 1988

One of the most original inventions to combine **electronics** and **optics** was conceived by the American Dana Anderson of Colorado University in Boulder; it is a holographic computer capable of making three-dimensional images. In principle, a **hologram** is an image made up by a **laser beam** from a given object. It possesses the peculiar property that if a hologram is broken, the entire image can be reconstituted from its fragments, since all the information is stored in any part of the hologram; moreover, the smaller the fragment is, the clearer the image. The Anderson invention is based on this property, and is a system of image projection from two or more holograms which are crossed by the same (amplified) laser beam, thus leading to the production of composite images.

Anderson's system functions in this way: a laser beam illuminates an object and thus produces the first hologram, which is recorded; then the image of a second object is used to make another hologram which is recorded on the same input medium as the first. If one tries to project only one of these two holograms by passing the laser beam over just one part of one of them, both are in fact obtained, but they are superimposed and transformed into a composite image, analogous to that of the superimposed memory. The second image is weaker.

Anderson's invention actually aims to construct an optical model for the study of **memory**, as well as phenomena such as association, forgetfulness and obsessions. By equipping a computer with the ability to record many holograms, to be used at random, the researcher hopes to provide a tool for the use of neurologists which will enable the greater understanding of such things as dreams, as well as the fading, or on the other hand, the reinforcement, of certain memorized images.

Interferometer

Young, 1802; Jamin, 1856; Fizeau, 1862; Michelson, 1881;
Rayleigh, 1896; Fabry and Pérot, 1897; Ryle, 1960;
Labeyrie, 1975

An interferometer is a device which uses **wave properties of light** to study the structure of objects. It belongs in the domain of optics and is capable of very high precision measurements of infinitesimal physical quantities, such as the length of the wave itself, and of large quantities, such as distances of 100 m which have to be very precisely measured, or infinitesimal quantities which represent significant distances, such as the **angular distance between two stars**.

The principle behind it is simple: the light rays from one source which are separated and then recombine form **interference fringes**, for example when a screen with two slits in it is placed in the path of the rays and then a second plain screen is placed behind the first. These fringes look like dark bands alternating with light bands on the plain screen. The number of bands depends on the apparent distance between the sources; at the lower limit, if the two slits are theoretically made to coincide, there is no more interference. But if an object of an equal size to the length of the light wave, that is, $1/2000$ mm, is placed in the path, then the same effect would be obtained as when there were two slits. If the object measures $4/2000$ mm, four interferences would be obtained, and so on.

The principle of optical interference was known and described (but without full understanding) in 1665 by Robert Boyle, Robert Hooke and Sir Isaac Newton; it was their compatriot Thomas Young who in 1802 explained it as the superposition of light waves. Young did not build the interferometer which carries his name but his work was essential to its development. The interference fringes were studied by numerous scientists in the first half of the 20th century, notably the Frenchman Augustin Fresnel and the German Wilhelm Haidinger, but the first to invent an instrument which made use of them was the Frenchman J.-C. Jamin who in 1856 built an **interference refractometer**. It was a device which measured the refractive index of a medium by studying interference fringes. This was based on the observation made in 1817 by Sir David Brewster which Haidinger studied more deeply in 1849: that when light is reflected by two parallel thick plates of glass which are slightly tilted, the fringes formed are equal and sloping. This device was to lead to the invention of the **surface interferometer**, first by Hippolyte Fizeau in 1862, then Louis Laurent in 1883; the device is still used in the optics industry for the verification of flat surfaces. In 1891 the Austrians Ernst Mach and L. A. Zehnder improved the Jamin interferometer and built one which was named after them for the study of laminar flow.

Fizeau's greatest contribution to interferometry was his invention of a method which enables the measurement of the angular separation of two stars. This was achieved by the superposition of the interference fringes produced by each of the two light sources after division by a slotted screen and it dates from 1868. It led to a major invention by the Englishman A. A. Michelson in 1881, the **stellar interferometer**. In 1890 Michelson managed to measure the diameter of the moons of Jupiter which measure no more than 1 arcsecond across. Michelson is also responsible for a slightly different interferometer which is especially famous because it allowed the experiment to take place which founded the **theory of relativity**.

Between 1881 and 1887 Michelson and his compatriot E. W. Morley prepared a series of measurements of the speed of light relative to the Earth moving through the cosmos, to study the speed of the Earth relative to its cosmic environment. When it took place in 1887 the experiment showed that the speed of light relative to Earth was essentially constant, and this led to Albert Einstein's formulation of the **Special Theory of Relativity** in 1905. Interferometry played a major role in the cosmology and physical theory of the 20th century. In 1896 the Englishman John William Rayleigh improved the Jamin interferometer by making a refractometer which had two achromatic lenses and a long focal length to measure the refractive index of fluids. It is now used under the name of the **Rayleigh–Haber–Löwe interferometer**.

One of the main contributions of interferometry at the end of the 19th century was, on the one hand, the definition of the **standard metre** in terms of wavelengths, and on the other, the precise definition of wavelengths. The latter was achieved using the **Fabry–Pérot interferometer** which was made in 1897 by the Frenchmen Charles Fabry and Alfred Pérot. It constituted significant improvement of **mirror interferometers**, such as the Mach–Zehnder, which make use of the fact that successive reflections of a light beam on mirrors produce very clear, fine interference fringes, which enable measurements as small as 1/600th of the wavelengths of visible light, or 1/1 2000 000 mm, to be taken.

The significance of Michelson's discovery encouraged him to continue with it. He formed a partnership with his fellow countryman, the astronomer Francis Gladhelm Pease, in order to build an interferometer which had openings 7-m apart; its theoretical power of resolution actually reached 0.02 arcseconds, that is, it was 50 times more powerful than the previous one. In this way in 1920 Pease measured the diameter of the star Betelgeuse in the constellation Orion.

In 1975 the French researcher Antoine Labeyrie made another major invention which consisted of an **interferometer** with two reflecting telescopes with 25-cm diameter mirrors which could be placed a distance of 6–35 m apart; this new apparatus, which required some strong technical constraints (it actually needed mechanical stability on a previously unknown scale), had a resolution of 0.003 arcseconds, that is, it could distinguish between two objects 1 m apart at a distance of 60 000 km. The interpretation of fringes observed in this way was processed by computer. This device led to the development of an even more powerful interferometer, using the same principle, which consisted of two telescopes with mirrors 1.5 m in diameter situated several hundred metres apart and this time capable of a resolution of 0.0002 arcseconds. The interferometer has been of great use to astronomy and enabled previously mysterious celestial bodies to be measured, such as double stars.

Radio-interferometry has been added to optical interferometry since the 1960s; it works on the same principle but radio waves are its means of exploration. It was the Englishman Martin Ryle who, in 1960 (at the Cambridge Radio Astronomy Observatory), directed the development of radio 'aperture synthesis'. This consists in doing interferometry with several radio telescopes of different separations to build up a detailed picture of an astronomical object. Radio-interferometry provides considerably finer measurements than optical interferometry. Similarly, **laser interferometry** has enabled miniscule measurements to be carried out since the 1970s.

Optical parametric oscillator
Collective, 1970–85

An optical parametric oscillator, also called O.P.O. for short, is a very fine measuring instrument which is meant for measuring physical phenomena which cannot be measured with **lasers**. It can be attributed to neither a particular inventor nor firm but is the result of the efforts of numerous physicists all over the world, notably in the United States and in France, in the AT&T Bell Laboratories and the Hertzian Spectroscopy Laboratory of the École Normale Supérieure in Paris.

Since 1970 the work was aimed at establishing a system for measuring minimal movements, such as those that could be communicated by **gravitational waves** to particles or molecules; these movements remained undetectable because of the slight fluctuations of a laser beam, known as **'quantum noise'**. The fluctuations in question are not due to any fault in the construction of the lasers but are in the very nature of the beam, the phases and intensities of which cannot be exactly determined — in quantum mechanics, the exact velocities and positions of photons can never be simultaneously determined. If, for example, one wants to reduce the phase quantum noise in a laser, then the intensity quantum noise must be increased, and vice versa, and some residual noise always remains.

The O.P.O. essentially consists of a **photon** emission system and the photons are, as it were, split by a crystal; the splitting of the crystal produces twin photons at exactly the same time. The two photons cross over the crystal again, where they stimulate the generation of new photons. The flux of photons would be too great if it were not for the fact that some escape through the system of mirrors; the live emission of photons is very weak and so devoid of quantum noise, but nevertheless strong enough to oscillate spontaneously like a laser. Since there is a double emission of photons, two intense directed beams are also created. This is a shortened account of the O.P.O. principle which has been modified by various laboratories following complex methods, all with the basic intention of completing work in **fine optical interferometry**.

Oscillograph
Braun, 1897

An oscillograph is an instrument which enables the observation and recording over time of variations of a physical property.

The first oscillograph of all was invented in 1897 by the German Karl Ferdinand Braun on the following lines: in a **cathode-ray tube** the electron beam follows a straight path. If the tube is placed in an **electrical field**, the beam undergoes a deviation which is proportional to the strength of the applied field. Braun, who held the Chair of Physics at Strasbourg University from 1895, built a device comprising a cathode-ray tube which would enable the study of high electrical frequencies.

This apparatus eventually led to the invention of the **television tube** via the **oscilloscope**, which is actually just an oscillograph with a screen. It must be noted that the terms 'oscillograph' and 'oscilloscope' are often used synonymously because their principles are identical.

Photogrammetry
Laussedat, 1846; IGN, 1972

Photogrammetry is the discipline of measurement of dimensions using two different photographs of the same object; it has substituted for topography and geometry and is currently used for establishing the outline of towns or public monuments or of objects too fragile to withstand direct contact.

Photogrammetry was discovered in 1846 by Aimé Laussedat who, after experimenting on the façade of the Hôtel des Invalides in Paris in 1849, carried out the first **topographical survey**, this time on the

Photogrammetric *plotting of the monument shown above.*

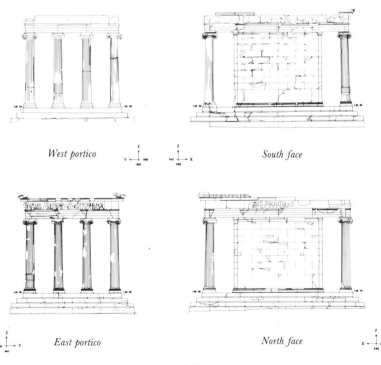

West portico

South face

East portico

North face

Chateau of Vincennes in 1850. 'Reinvented' and completed by the National Geographical Institute in 1972, it requires photographic instruments, with optically and geometrically precise features, which are also called **metrical chambers**. Given that the position of a point on the negative must be measured to the nearest 0.01 mm, the back of the chamber and the photographic plate must be strictly level; the optics must be totally devoid of sphericity and deformations — conditions for very high definition or **orthoscopy**.

Photogrammetry is also used for the surveillance of works of art, bridges and dams, as well as for checking the accuracy of the parabolic reflectors of large radio telescopes which are subject to deformation (see illustration p. 149).

Until the discovery of photogrammetry, in order to draw up the plans of an existing building, measurements of the actual building had to be taken and then recorded on paper. This was often a risky undertaking. Afterwards it was possible to carry out the same surveys by photography, as in the case of the Greek monument, the Apter Victory Temple on the Acropolis, of which the exact photogrammetric notes are shown.

Photographic lens

Goddard and Sutton, Harrison and Schnitzer, Ross, Busch, 1860–5; Steinheil, Dallmeyer, 1866; Abbe and Schott, 1888; Clark, 1889; Rudolph and von Hoëgh, 1893

After 1850 the introduction of **wet collodion plates** enabled indoor photographs to be taken and it was soon observed that all the existing lenses caused irritating distortions. The Englishmen J. T. Goddard and T. Sutton began to put this matter right by modifying the Petzval lens, but it was the Americans C. C. Harrison and J. Schnitzer as well as the Englishman Thomas Ross who in 1860, almost simultaneously (followed in 1865 by the German Ernst Busch) achieved the best improvement by inventing a **biconvex lens** of which each section had the same radius of curvature (**globe lens**). The distortion disappeared, but both the spherical aberration and the peripheral halo remained.

In 1866 the German A. Steinheil and the Englishman J. H. Dallmeyer independently invented a type of objective lens with double elements exactly symmetrical either side of the diaphragm. This had the advantage of causing the halo to disappear. Moreover, the external shape of the lens 'flattened' the field and the spherical aberration disappeared. Both Steinheil and Dallmeyer achieved an image which was hardly distorted at all for numerical apertures as fast as $f/7$. This objective lens was the best available until the beginning of the 20th century.

The quality of the actual glass still needed to be improved. It had been the practice to use crown-glass for the positive lens and flint-glass for the negative lens. In 1888 the Germans Ernst Abbe and Otto Schott made **barium glass** which, like the crown-glass, had weak dispersion but also had a high refractive index. This enabled the manufacture of positive non-divergent and convergent lenses, thus achromatic lenses.

Two years later the German Paul Rudolph, who was working for the firm Zeiss, redesigned the Dallmeyer objective on the following basis: he made a double objective which had one lens divergent and concave to the diaphragm and the other convergent and convex to the diaphragm. This resulted in the complete annulment of both the astigmatism and the spherical aberration (**Protar Zeiss**).

There was new progress in 1893 which was achieved at once by the same Rudolph and also von Hoëgh from the firm Goerz. This was the objective with six lenses fixed

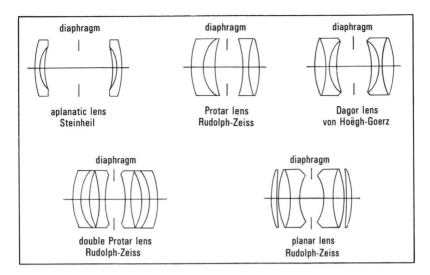

Photographic lenses.

together in threes with refraction indexes crossing from the diaphragm (1.52, 1.57 and 1.62). The field was increased and the astigmatic correction improved, but only with the apertures $f/6.3$ and $f/8$ (**Triple-Protar Zeiss** and **Dagor Goerz**). From this time onwards improvements came in quick succession.

In 1817 while studying telescope objectives the mathematician C. F. Gauss had discovered that a particular lens disposition and design could lend the same spherical aberration to any length of light wave, and so eliminate **sphero-chromatism**. In applying this discovery to photographic lenses, in 1889 the American Alvan Clark registered a patent which showed that by varying the space between two lenses of this type on either side of the diaphragm, with the concave surfaces facing each other, the curvature of the field could be varied at will. Rudolph, who had had the same idea, noticed that it was however impossible to

eliminate the chromatism in this way. He tried using a material of maximal diffusing power for the **adhesive interface** between the lenses (for, beginning with the Steinheil objective, the different shaped lenses were stuck together using adhesives which varied in type depending on the inventor). Thus he made a type of objective which offered the user much greater adaptability (**Planar Zeiss**, 1895).

Research into objectives was to continue right through the 20th century, helped particularly by the use of rare earths in the preparation of crystal. In 1950, **lanthanum glass** enabled the manufacture of objectives with minimal margins of chromatism. During the 1960s research into making it lighter was to lead to the objectives being made of **polymethyl methacrylate** for the lenses of several lower range camera parts, which had the concave element made out of a high dispersion plastic such as **styrene**.

Photographic photoelectric cell
Rhamstine, 1931

In 1931 the American J. Thomas Rhamstine thought of using the principle of the photoelectric cell in **photography** with the aim of measuring the **luminosity** of the object to be photographed and adapting both the time of exposure and the aperture of the lens to the sensitivity of the film. The photographic photoelectric cell was commercialized under the name of Electrophot.

Positron microscope
Michigan University, 1987

This variation on the **electron microscope** (see p. 144) is based on the fact that positrons, the symmetrical particles to electrons, react with matter in a different way to electrons and therefore give a different and complementary image of the material under study to that obtained with the electron microscope. Invented in 1987 by James C. Van House and Arthur Rich of Michigan University, this microscope uses positrons produced by **isotopes** such as **sodium-22**; it also inspired the same inventors to make the **positron re-emission microscope** which uses the effect of the absorption and re-emission of a fraction of these particles.

Quartz clock
Marrison, 1929

The quartz clock, ancestor of modern quartz watches, was invented in 1929 by the American clockmaker Warren Alvin Marrison who took account of the following facts: when a crystal of quartz changes shape its surfaces become oppositely charged (resulting in an electric field across the crystal), and inversely, an applied electric field causes deformation of the crystal. This property, defined as the **piezoelectric effect**, enables the **frequency of alternating electric current in a circuit** to be controlled using a crystal of quartz.

The first quartz clocks were not very reliable for they tended to accelerate suddenly or stop without warning. It was only 30 years after they were invented that improvements could be made.

Radar

Tesla, 1900; Hülsemeyer, 1904; Marconi, Taylor, Young, 1922; Watt, 1936

The radar is an instrument for **detection** based on the principle of **electromagnetic wave reflection**. In principle it consists of locating an object at a distance and studying the shape of it on the screen of an **oscilloscope** which receives the **echoes** of the reflected wave beam. The radar is derived from an observation made in 1866 by the German Heinrich Hertz: that electromagnetic waves can be reflected, refracted and diffracted. Hertz made this observation when studying radiation of the same wavelength as that of modern radar waves, about 60 cm, but he did not see its potential application.

It was in 1900 that the Croat Nikola Tesla was the first to describe the possibility of locating a moving object using **radio wave** echoes. Technology then was insufficient to guarantee this detection system which was evidently destined in the first instance for ships, a point which would lend it some practical significance. In 1904 the German Christian Hülsemeyer presented the patents of a radio detector based on the same principle in several countries. Its purpose was to prevent collisions, but this detector never reached industrial production stage. In 1922 the Italian Guglielmo Marconi and the Americans A. H. Taylor and L. C. Young took up the principle again just as it was but with the improvement of a simple method of localization: when the propagation speed of radio waves was known, all that had to be done to determine the distance at which the object was situated was to halve the time which elapsed between emission and reception of the echo from it. This method involved detection by 'pulse waves', the instant of emission of which was much easier to know than that of continuous 'wave trains' which had been used until then.

The pulse waves were first used by the Americans Gregory Breit and Merle A. Tuve when they were measuring the altitude of the ionosphere, the atmospheric layer of ionized gases which reflects radio waves; since then only pulse waves have been used in research on radar (which was only so-called in 1940). Many industrial countries were carrying out this kind of research, not only for reinforcing navigation safety but also for blatantly military reasons. The research was difficult and was punctuated with numerous failures mainly

The British and American progress in radar research during the decade 1935–45 was helped considerably by an excellent **centralization of research**, whereas Germany lagged behind in this area because of Hitler's total lack of scientific education. In 1935 Germany was at least equal with the United Kingdom: the first German radar, the 125 MHz Freya, which was controlled by the maritime signal office under the supervision of Rudolf Kühnold, produced promising results; these were confirmed by the 560 MHz radar built in 1938 by the firm Telefunken and the Luftwaffe at Würzburg. Over eight military research centres and 200 German institutes were working on improving the radar, under the somewhat vague direction of Göring. In 1940 however, Hitler banned all electronic research, however fundamental it was, under the pretext that electronics was a 'Jewish science'. Work was only resumed again in 1943, when the Germans shot down a British aeroplane which was transporting **high-resolution short-wave radar** equipment, which explained how the Allies had been so successful in their fight against the submarines. But it was too late and when Hitler decided to call up the 6000 scientists who were in his forces in order to recommence research in electronics, 2000 of them were absent from the roll call. Similar confusion put Japanese research at a disadvantage, as it did with Soviet research, although these three totalitarian states had boasted that they would win by good organization.

for one reason: the radar only works if the initial beam is strongly focused, otherwise the echo thrown back from the surface of the water or the earth may overshadow the much weaker echo from the target. Satisfactory **focalization** happened only in 1936, when the Englishman R. A. Watson-Watt thought of amplifying the high-voltage pulses with a **magnetron** (a two-electrode device with electric and magnetic fields) and of using an amplifier for the echoes, with the emission being concentrated by an electron gun installed in the cathode emission tube. This **Radio Detector Telemeter** enabled a chain of protective stations to be built along the British coast.

An improved American version, the **SCR-270**, was made in the United States in 1938. The radar was tremendously useful to Great Britain in 1940 and then to the whole of the Allied Forces as the airforce were enabled to fly in poor visibility as well as to carry out precision bombing. Naval defences could detect convoys and German submarines (using radar in 1943, about one German submarine every day could be run down); aerial defences could recognize national aeroplanes which were equipped with **transponders** which sent out coded messages automatically on reception of the radar signal.

Radiometer

Melloni, 1850; Langley, 1878; Boys, 1888; Ångström, 1893

A radiometer is a **device for measuring radiation**, but the term is reserved for the measurement of the heating power of radiation. Traditionally it includes neither the selenium cell, which is sensitive to visible light (and produces electricity), nor the photoelectric cell, which is sensitive to actinic rays (which have a chemical effect on certain substances), because they are relatively insensitive to infra-red radiation and so they do not measure energy reliably for all wavelengths of radiation. The radiometer has played a fundamental part in physics, thermodynamics, electricity, optics and in numerous disciplines such as the study of the Sun. Not including the thermometer, the first ever radiometer was the **thermopile** invented by the Italian Macedonio Melloni in 1850. It was a set of a hundred electric 'thermocouples' assembled in the shape of a cube, with each thermocouple consisting of two elements, one made of an alloy of antimony, the other of an alloy of bismuth. The sensitivity of these was such that a 1°C temperature difference between the cold element and the hot element would generate a 120 microvolt potential in each couple, with the whole thing producing an electromotive force or voltage, of 120 millivolts. The device was

inconvenient in that it was slow, because of its large thermal capacity (i.e. a large amount of energy was required to cause an appreciable temperature difference).

In 1878, the Englishman S. P. Langley built a radiometer of his own invention which was much quicker. It was made of two wire meshes painted black and of equal resistance which were balanced on a kind of construction called a 'Wheatstone bridge'; when a current was induced, the one which was exposed to radiation heated up, and the other did not; the temperature difference between the two produced a difference in current which was measured by a galvanometer. This was better than the thermopile for as the deflection of the galvanometer was heightened, the increased electric currents could be recorded. Radiometers of this kind which were very fast and quick, called **bolometers**, were made using very thin receivers. The problem with them was that as the deflection of the galvanometer increased the current itself caused the meshes to heat up making the device no longer reliable.

In 1888 the mathematician and physician Charles Vernon Boys invented a **radio-micrometer** which combined the radiometer and the galvanometer, and which was fast, sensitive and constant. It was a delicate instrument, consisting of a very light thermocouple attached to a wire ring made of copper, itself suspended by a quartz thread between the poles of a strong magnet; it could detect even weak radiation.

Interest began to be shown in the **absolute measurement** of radiation and no longer only in the estimation of radiation from a source at a known temperature. During the 1880s the Frenchman C. S. M. Pouillet invented an instrument, the sensitive surface of which was comparable to that of the mesh of the bolometer: it was a black disc of a given thermal capacity, the heating of which in a given time could be measured. As it was inconveniently slow, and could not take account of the heat loss taking place during heating, it was improved by several physicists (G. G. Stokes, J. Violle and A. Crova) in the following way: the sensitive surface was made of a disc of low thermal capacity and high conductivity, which was shut into a sealed cylinder and maintained at a constant temperature. This was the **pyrheliometer**, which is still used for the **measurement of solar radiation**, but which is not suitable for the study of weak radiation.

Just before the end of the 19th century the famous Swedish physicist Anders Jonas Ångström invented a new type of pyrheliometer. This contained a **manganin** (an alloy of copper, manganese and nickel) acting as the sensitive surface, which was placed in front of a captor of the same size but which was heated by an electric current in a way strictly equal to its counterpart, using a **rheostat** (enabling the current to be varied by varying the electrical resistance). The amount of current necessary to heat a captor indicates, by analogy, the amount of captured radiant energy.

Raman effect spectometry

Raman, 1928

This technique is an instrument for the extremely precise analysis of **polyatomic molecules**, based on an effect named after the man who discovered it, the Indian C. V. Raman. In 1928 Raman discovered a **frequency variation in the diffraction of light** which strikes a molecule of gas, liquid or solid; it takes place according to an incidental monochromatic light radiation. In general, this monochromatic light is **infra-red**. The phenomenon is explained by the fact that the dynamism of the molecule under the effect of the radiation changes its polarization, either because the radiation makes it vibrate, or because it makes it turn around. This is a necessarily vague explanation, for at the end of the 1980s the phenomenon was still being studied. Raman had the idea of applying his discovery to the analysis of minuscule quantities of any substance, the nature and structure of which the phenomenon would be able to define since it received the information from the vibratory frequencies of the component atoms. Indeed, when the monochromatic light is diffracted it forms **specific spectra** of the substances which it strikes.

Raman spectroscopy proved to be irreplaceable in physical and chemical analyses since neither ordinary spectrometry nor gas chromatograph could equal it in sharpness of definition.

'Realistic' cartographic projection
Van den Grinten, 1904; Peters, 1974; Robinson, 1988

The main problem with all systems of representing the Earth on a flat surface is the distortion of the proportions. On a 'classic' cylindrical projection such as the Mercator for example, Greenland is 16 times larger than in reality and seems to be even bigger than South America. In 1904 the American Van den Grinten devised a system called moderated aphylactic which was adopted by numerous bodies, but which nevertheless included considerable distortions. Even if the polar regions in it were less significant than in the Mercator projection, they still 'benefited' from abnormal enlargements; Alaska thus featured as an area multiplied five times and equal in size to that of Brazil, for example, which in reality is six times larger. In 1974 the German Arno Peters completed a method which more or less re-established the proportions of the salient land masses and the oceans between them, but which greatly deformed their outlines. In 1988 the American Arthur Robinson improved the Van den Grinten system again by obtaining both correct and 'realistic' proportions and outlines. However, as acknowledged by the inventor of the method, it is still an 'artistic' adaptation which itself contains numerous distortions.

As there is no way of projecting a sphere onto a flat surface without some distortion, it is likely that Robinson's invention will not be the last in cartography.

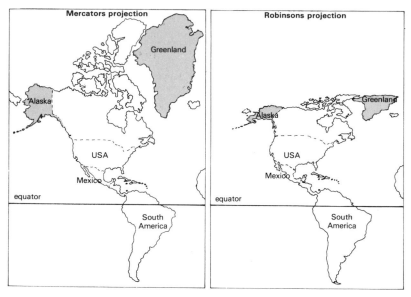

*All cartographic systems inevitably include distortions since they consist of projecting spherical images onto a flat surface. The scale of the distortions in a **Mercator projection** is obvious: Alaska appears larger than Brazil while it is in fact nearer to the size of Mexico, and Greenland appears considerably larger than South America which is not the actual case. In the **Robinson System** the projection is noticeably closer to the truth although not exact.*

Richter seismographic scale

Richter, 1935

The seismographic Richter scale is for measuring the size of **earthquakes**. It is an open-ended logarithmic scale where the progression of degrees corresponds to the increase in the effects of the earthquake. Level 1 would indicate an earthquake which at 100 km from the epicentre would move the needle of a seismograph a tenth of a millimetre for 0.8 s; it might not be noticed by people in the area. Level 9 would be exceptionally violent. This scale was invented in 1935 by the American Charles Francis Richter and has been universally adopted since then.

Scanning tunnelling microscope (STM)

Binnig and Rohrer, IBM, 1982

The scanning tunnelling microscope is a considerable improvement on the **electron microscope** (see p. 144) and is based on the following principle: two segments of electric wire can conduct an electric current if their tips are near enough. Usually, requirement of continuity across the gap which separates the two segments does not allow the electrons which have flowed through the first segment to continue their course in the second, so current cannot flow. **Quantum mechanics** postulates — and in this case shows — that atomic particles cannot be defined just as discrete quantities; they are also **waves**. If the distance between the two segments is greater than a certain, calculable amount, the wave disappears; if this distance is reduced sufficiently, the energy of the wave allows it to be crossed. The electrons then behave as though they had made a tunnel through the obstacle which was hindering them.

The above picture is meant to give an idea of what an ultra-powerful modern microscope is like: it is the scanning tunnelling microscope which can be held in the palm of the hand and does not look anything like a traditional microscope.

In 1982 the Swiss Gerd Binnig and Heinrich Rohrer from the IBM laboratory in Zurich invented a microscope which made use of the tunnel effect. The principle behind it is simple: a very fine needle which is conducting a current is placed very close to the object to be studied. Since the intensity of the current transmitted to the object depends on the distance from the needle (it varies exponentially according to the distance separating the electrodes), it will increase when the needle passes over a protuberance and will diminish above a hollow. It is therefore possible to establish in a very precise way the differences in level or relief of the object under study. The precision obtained is 0.1 Å (1 Å = 1 Ångström = 1 ten millionth of a millimetre) to the vertical and 1 Å horizontally, approximately the size of an atom; consequently an image of the surface of an object is obtained which is on an atomic rather than just a submicroscopic scale as it was with the electron microscope. At the AT & T

Bell Laboratories a single atom has been placed onto a germanium surface.

This instrument proves to be incomparably useful in the study of the surfaces of materials used in electronics, such as silicon, a free layer of which is known to reorganize itself in a different way from the internal layers of the crystal. The scanning tunnelling microscope also enables the shape and size of a **virus** or **DNA** to be measured.

The making of this instrument incurred some very high level technical problems to do with the stability of the electrode point (which when made of **tungsten** has only a single atom at its tip). This had to remain fixed at a few Å from the object under study and be totally free from environmental vibrations, including sound vibrations. The prototype consisted of a stage which is suspended by magnetic levitation; the movement of the needle is controlled by a piezo-electric element.

Speaking clock
Sivan, 1895; Esclangon, 1932

By combining the recent invention of the gramophone with clock-making, Sivan, a Swiss manufacturer from Geneva, made the first known speaking clock. It consisted of phonographic recordings which were set to play on the hour. The idea and the

making of the speaking clock for the telegraph service are attributed to the celebrated scientist Ernest Esclangon, who was then director of the Paris Observatory. It was a recording synchronized with a clock.

Stereoscopy
Ducos du Hauron, 1893

The first stereoscopic, or three-dimensional, images were made by the Frenchman Louis Ducos du Hauron in 1893. The inventor's own explanation is perfectly clear: 'The characteristic of the process consists of the way that shadows and black are made;

they are produced not by a black pigment [...] but by the crossing [...] of two colours one of which intercepts the other. With this interception being translated at each point into black which is in proportion to the intensity of the intercepted colour, a phenomenon of "antichromatism" takes place which is analogous to that which, in the system of "pigmentary heliochromy" of which I am the inventor, recreates the natural black by the superimposition of colours. Example: when the image which corresponds to the perspective of the right eye is printed in red (red lead or vermilion) on a white background, and the image corresponding to the perspective of the left eye is printed in transparent violet-blue (Oriental blue) overlapping the red image, if a violet-blue (Cobalt blue glass) dioptric (refracting) medium (glass or film) is inserted between the double, mixed image constituted as described above and the right eye, and a red medium (such as ruby-red glass) is inserted between this same double image and the left eye, a very curious result is achieved: (1) each eye will perceive in black an image, the colour of which does not correspond to that of the interposed filter; (2) neither the left eye nor the right eye will see the image of a colour corresponding to that of the filter used for each of them. Concerning the second part of the phenomenon, the explanation is that [. . .] for the right eye the blue disappears while the red is translated into black. Inversely, the red disappears for the left eye which is armed with red glass, whereas the blue is translated into black.'

Ducos du Hauron appears to be the precursor of two discoveries. The first, the anatomical **crossing of optic nerve fibres**, was to explain in the 20th century that vision is the product of retinal sensations from both eyes which are superimposed and transposed. The second, **holography**, is based on the laser decoding of the different phases of light waves on an object also illuminated by the laser.

Vibration analyser

Ometron, 1987

There are numerous techniques for the **analysis of vibrations in materials** which are to blame for stress and also for cracks. An ultra-sensitive instrument called the Vibration Pattern Imager was patented in 1987 by the American firm Ometron Inc., based on **laser interferometry**. It is capable of recording vibrations as slight as a fraction of the wavelength of light and works on the basis of an amplification of the differences between the source of emission and the reflection of the laser beam, or the **optical interference pattern**, when the laser is directed onto the vibrating material.

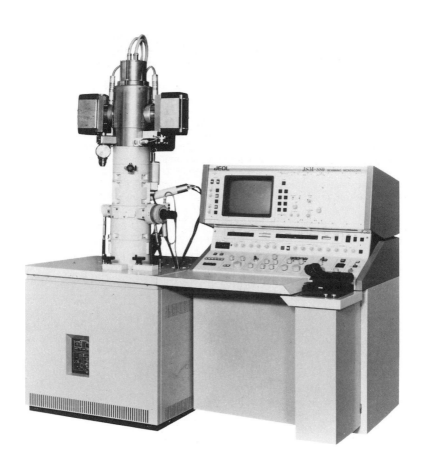

Electron microscope which works by
scanning EOl (JSM 890) and it equipped with
a field emission gun. This apparatus is the first to
achieve a resolution of 0.7 nanometres.

Medicine and Biology

The appearance of inventions in the fields of medicine and biology is not in proportion to the development of these two sciences. These inventions seem of secondary importance compared to the tremendous changes brought about by the progress in the knowledge of living beings.

Until about 1950, to pick a landmark date, medicine and biology had obeyed the old, almost mechanically categorical ideas, which were inherited from classical eras. A certain organism suffered from a certain symptom, so it was given a certain product meant to treat the cause of the symptom. The remedy was indispensable, for one of the doctor's duties is to prevent the patient's suffering. But it was rather rough-and-ready; often it did as much harm as good. From around 1950 the interpretation of a live system began to change from the static to the dynamic, and gradually it was understood that the equilibrium which controls life is based on a multitude of sub-system interactions. Attempts were made to discover what these sub-systems were: from the tissue to the cell and from the cell to the infinitesimal, such as DNA. During the 1960s the 'old' biology was examined in the light of molecular biology; there is really no longer any other type of biology.

In consequence, inventions which made sense only in this context began to take place, such as products based on monoclonal antibodies, which are themselves imitations of the products of the immune system, the genetic transplant which was unthinkable 30 years ago, and the transgenic mouse which was unthinkable only 10 years ago. They are fundamental concepts which are already changing the course and the future of medicine, whilst their practical consequences have hardly been fully exploited.

Inventions in surgery also began, such as the gamma helmet and the bistoury laser, which reduce the art of using the scalpel at the beginning of the century to a kind of hopeless butchery, and medicines which are to the old pharmacopoeia what penicillin was to the herbs of the apothecaries.

In this particular domain inventions seem to relegate the last thousand years of medicine and biology to prehistory.

Artificial blood
Clark and Gollan, 1966; Sloviter, 1967; Naito, 1979

Research on a blood substitute which can assume at least one of the blood's functions, **oxygen transportation**, seems to go back to 1933 but it was not conclusive then. It was only in 1966 that the Americans Clark and Gollan managed to keep some mice alive for a few hours, immersed in a liquid which flooded their lungs and ought to have led to their death. This liquid was an emulsion of a **fluorocarbon** in water. The fluorocarbon molecules possess the property of being able to link up with significant quantities of oxygen, provided that there is an abundance of it in the surrounding material; in the case of the mice, the oxygen was extracted from the water. These were the beginnings of the invention of blood substitute. In 1967 the American Henry A. Sloviter made the most of this experiment by injecting some rabbits with a fluorocarbon emulsion, with **physiological liquid** and some **albumen** (egg white). He based the theory behind this mixture on the fact that fluorocarbon alone does not mix with blood, and he claimed that the survival of

the animals was linked to the proportion of the injected emulsion in relation to the total volume of blood: if this was above a third the animals would die, because the substituted liquid would not be enough to ensure the transportation of oxygen and carbon dioxide. Two years later the American Robert Geyer improved the formula of the emulsion and achieved the survival of a rat after a total transfusion of blood. The first experiment on man was carried out by the Japanese Ryochi Naito who injected himself with 200 ml of **Fluosol DA**, milky-looking artificial blood, which he had made. Since then several other similar formulae which are more compatible with human physiology have been put forward. Whatever the situation, artificial blood is only used in conjunction with a transfusion of real blood in some emergency cases, such as the primary treatment of **third-degree burns**, for since it does not capture enough oxygen at lung level and does not release enough into the tissues, it requires respiratory assistance.

Artificial heart
Demikhov, 1937; Kolff, 1958; Jarvik, 1965–80

The first entirely artificial heart to be implanted into an animal was in a dog in 1937 by the Soviet researcher Vladimir P. Demikhov. Three successive attempts at implanting a rotary pump inside the thoracic cavity gave results which were satisfactory enough for Demikhov to continue his research until 1958, when he abandoned them. The next attempt was that of the American Willem Kolff, the inventor of the **artificial kidney** (see p. 165). The calf which had received the implanted heart survived for about 90 minutes. After several improvements following further experiments using calves, a calf survived for 121 days in 1969. In December 1982 Kolff

implanted an artificial heart in a man, Barney Clark, who survived 112 hours. This heart had been developed by the American Robert Jarvik; made of aluminium and polyurethane, it was the seventh artificial heart to have been made by this prosthetist, after whom it was named **Jarvik-7**. It consisted of a double pump which carried out the functions of the two natural ventricles (heart chambers); it was not automatic, for it had to be connected externally to a bulky (150 kg) piece of equipment. Jarvik had begun his first experiments in 1965. He continued with them until the end of the 1980s. At this time the artificial hearts (of which the Jarvik-7 and the **Penn State**, made by the University of Pennsylvania, were the only commercialized ones) had too many risks and shortcomings to be considered as definitive prostheses. Over and above the need to be connected, which implied a great restriction to freedom of movement, this type of apparatus came with the risk of **thrombosis** (blood clots), the formation of **foreign connective tissues, haemolysis** or the alteration of red blood cells, and malfunction due to the **rupture of diaphragms**; therefore it is considered only as a transition instrument during the waiting period for a biological heart.

The progress made in the domain of **miniaturization**, that is, of **batteries** and **accumulators** sufficiently small and light to be incorporated into an artificial organ, and sufficiently powerful to last several years, means that automatic hearts may be possible by the end of the 20th century. Progress in this sphere has been made in France, Japan, Germany and the United States among other countries. However, the state of knowledge at the end of the 1980s reflects the fundamental problems of **compatibility** between artificial and organic tissues, and remain unresolved. Some research suggested making artificial hearts with walls covered with living cells obtained by culture, and so covering over the surfaces of the artificial organ.

The hand of Robert Jarvik holding the most recent prototype of his **artificial heart**, *the Jarvik-7.*

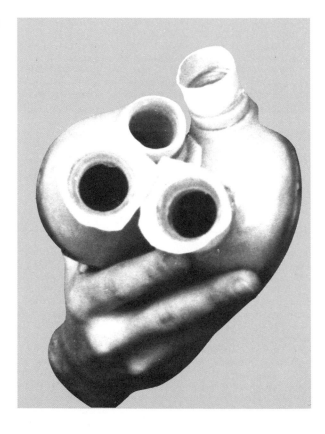

Artificial kidney

Abel, Rowntree, Turner, 1913; Kolff, 1945

The first artificial kidney ever made was the one invented by the Americans John J. Abel, L. G. Rowntree and B. B. Turner in 1913; it consisted of a set of porous tubes submerged in a colloidal solution. The object of it was essentially to demonstrate the **detoxification effect of the kidney**, and the toxins in canine blood were actually removed when it was through these filters. However it did not remain as an experimental apparatus for long. Before it could be used on a human being, filters had to be found which this time could filter natural toxins, and also an efficient anticoagulant

which would enable the blood to circulate outside the body without clotting. **Cellophane** proved to be a reliable filtering medium and **heparin** an equally reliable anticoagulant. Using these discoveries as a base, in 1945 the Dutchman Willem Kolff made the first artificial kidney which could be used for purifying the blood of a human being suffering from renal insufficiency. This first apparatus was very bulky. Over the years the size has been reduced in such a way that the patients can use mobile units which are not cumbersome at all and carry out the **dialysis** at home.

Artificial skin

Tanner, Bell, Neveu, Yannas, Green, Thivolet, c.1980

Neither a precise date nor a single inventor can be assigned to the invention of artificial skin, which constitutes one of the most significant in the field of modern **prosthetics**. The invention, or rather inventions of artificial skin, for there are several variations, are descended from the research into the treatment of **third-degree burns** and are essentially dependent on the development of new materials which are compatible with the human organism, and which do not meet with rejection. They are also a result of the techniques of **live tissue culture** which appeared at the end of the 1960s. Indeed, in 1964 the American John Tanner treated some samples of natural skin, taken from the patient to be healed, by putting them between two rollers, in order to stretch them and confer upon them some

kind of mechanical homogeneity. Subsequently, cultures of living cells were created (by John Burke) in plates which could be grafted onto the burns. Very briefly, this method, which was improved by the American Bell and the Frenchman Neveu, consisted of inserting **fibroblasts** (connective tissue cells) into a support tissue based on **collagen** (a protein); this method was succeeded by one which advocated the replacement of the collagen (in fact, bovine collagen) by a synthetic material, a porous **polymer** derived from this collagen. It was combined with other substances which were also compatible with the human organism (chondroitin 6-sulphate, a polysaccharide extracted from shark cartilage). One of the pioneers in this field is Ioannis Yannas, an American of Greek origin who in 1981 succeeded in grafting this tissue onto third-degree burns and obtained significant results. This tissue, called **stage 1 skin**, provides protection against infection and allows the mother cells of the skin, the fibroblasts of the adjoining underlying tissue, to migrate to the surface and penetrate the artificial skin. Meanwhile the polymer

Artificial skin has paved the way to research on other tissues to be used as prostheses, for example in the pancreas. Thus, during the 1980s the Canadian Anthony Sun studied an **artificial pancreas** cultured with **beta cells**, which make **insulin**.

has been broken down by the patient's immune system and on this previously half-reconstituted terrain it became possible to proceed to superficial epidermal autografts which would 'take' much more easily than before. At the end of the 1980s the manufacture of stage 2 skin was in sight which would be cultured with skin cells (Howard Green, Harvard; Jacques Thivolet, Lyons Faculty of Medicine).

Bacterial test
Gram, 1884

The final quarter of the 19th century coincided with a rapid development in **microbiology**. Already in 1875 the German Carl Weigart had discovered that killed bacteria absorbed a specific dye called **picrocarmine**; this he did using a technique known since 1865 for colouring tissue sections for microscopic examination. His discovery was a very useful method for outlining the cellular structure of sections, thus making study more manageable. Other dyes were sought for and found, such as methylene blue, fuchsin and crystal violet. In 1884 the Dane Hans Christian Gram made both a discovery and an invention in this domain. The discovery was that when the bacteria were treated in a certain way, some absorbed the dyes and others did not. So he went on to invent a microbiological test, which is still used nowadays; this is the classification of bacteria into **Gram-negative** and **Gram-positive**, based on the absorption of crystal violet. This test has been extended to the identification of other micro-organisms. In this way it is known that **yeasts** are Gram-positive and **Rickettsiae**, Gram-negative.

Barbiturates
Fischer and von Behring, 1903

Barbiturates are a group of medications derived from **barbituric acid**; they depress the central and hypnotic nervous system, being distinctly **sedatives** and **tranquillizers** (see p. 184) and date back to the synthesis of the acid which was made in 1903 by the Germans Emil Hermann Fischer and Emil Adolf von Behring from the Marburg Institute of Hygiene. This synthesis was from ethyl malonate and urea, hence the name **malonylurea**. Barbituric acid itself has no hypnotic power but its derivatives do. It was Fischer and Behring who carried out the first experiments on the derivatives of barbituric acid, such as Barbital and Veronal. Barbituric acid had been discovered by their compatriot Adolf von Bäyer in 1864 whose pupil Fischer had been. Fischer was a brilliant experimental organic chemist, a domain where he had made his mark, and he achieved the synthesis of barbituric acid during his research on **purines**, a group of heterocyclic compounds. Both Fischer and von Behring were presented with Nobel prizes, but that of von Behring was for his work on serotherapy by antitoxins (treatment by injection of antibodies) of which he was the inventor, and that of Fischer for research in organic chemistry.

Biodegradable polymer
CCA Biochem, 1988

Polyactide was made by the Dutch firm CCA Biochem of Gorinchem for use in surgery. It is **polymerized lactic acid**, lactic acid being a natural product released by some micro-organisms during the fermentation of glucose. It comes in the form of **suturing threads, bone platelets**, or **artificial skin** and is gradually broken down by the effect of the human metabolism. In 1988 it was thought that it could be used as an envelope for **depot (delayed-action) medicines**.

Bistoury laser
Bessis, 1962; Ingram, 1964; Herriott, Gordon, Hale and Gromnos, 1967; Coscas, 1978

The possibility of using a laser for surgical purposes had been considered since the end of the 1950s, but it posed some delicate technical problems. The path was cleared in 1962 by the Frenchman Marcel Bessis and a group of biologists who made a micro-ray laser of 2.5 microns in diameter. In 1964 the American H. Vernon Ingram

A few years ago the bistoury laser was an avant-garde instrument; it has now become commonplace. The scene below, where an ophthalmologist is following the path of his ultra-fine beam on a screen, is common in all well-equipped ophthalmological clinics.

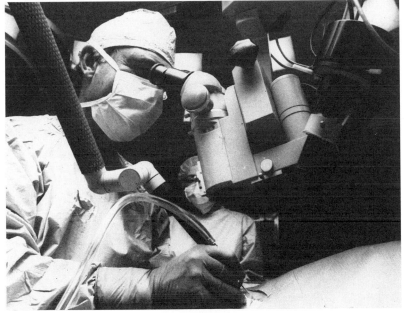

carried out the first surgical operation in **ophthalmology** (eye disorders) with a Nelas laser. The bistoury laser (a bistoury is a type of surgical knife) was then developed in 1967 by the Americans D. R. Herriott, E. I. Gordon, H. A. S. Hale and W. Gromnos from the American company Bell Telephone. It was these men who showed that such an instrument could cauterize and cut simultaneously. The first medical use of an argon laser was in ophthalmology, carried out by the Frenchman Gabriel J. Coscas in 1978.

Blood bank
Yukin, 1931

In 1931 the Professor Sergei S. Yukin, from the Moscow Sklifosovsky Institute, had the idea of keeping a stock of human blood stored under special conditions. This was to use in **transfusions** in the emergency service of this Institute. The actual expression **'blood bank'** was coined in 1937 by the American Bernard Fantus from Chicago.

Cardiac stimulator (pacemaker)
Hyman, 1930; Callaghan and Hopps, 1952; Greatbach, 1956

Electrical stimulation of the failing heart was used very occasionally in the 19th century but for a long time it remained totally empirical, for two reasons: the right type of current (continuous or alternating) and intensities to use were known; also it was not known which places, circuits or tissues to stimulate. It was known (since the work by the Irishman Robert Adams and the Englishman William Stokes in 1854) that a pulse of contractions exists between the ventricles and the auricles (heart chambers) but neither the nature of the automatic contractions of the cardiac muscle, or **myocardium**, nor the structure responsible for it were known; this structure only became conspicuous in 1906, when the Japanese Sunao Tawara described a node in the tissue of the myocardium, between the ventricles and auricles, which thenceforth would be called **Tawara's node**. The following year the Britons Arthur Keith and Martin Flack discovered another node which conveys and regulates the contractions of the entire myocardium, and which was named the **Keith and Flack node**. One would have thought that this would be the place upon which to act, in principle. In 1903 the invention of the **electrocardiograph** (see p. 172) provided an answer to the first question: the nature of the cardiac contractions is electric. The evidence took a long time to be accepted, for the inventor, Einthoven, only received his well-deserved Nobel prize for it 21 years later. But an important conclusion had been reached: it definitely is an electric current which must be applied to make a faulty heart work again.

The means of applying this current seem to have been found in 1898 by the French veterinary surgeon Auguste Chauveau; he introduced a probe, the point of which would stimulate the myocardium, into the heart via an artery or vein. This is what is called a **catheter**. However it is a danger-

> The first pacemaker to be implanted broke down after a few hours and was replaced immediately. Its beneficiary, Linsson, received his twenty-third pacemaker at 72 years of age.

ous process. In 1929 the American Gould proposed to replace it by the direct insertion of an **electrode-needle** into the heart across the thorax. In 1930 another American, Alfred Hyman, thought that the electrode-needle, by the trauma provoked, induced a difference in potential which was the same as an electric charge and that it would be more 'simple' to apply a modulated electric current to the Keith and Flack node. Its simplicity was not obvious, for the first cardiac stimulator ever built, that of Hyman, called the **Hymanotor**, weighed more than 7 kg . . . and its generator was started using a handle. Nevertheless the Hymanotor saved dozens of lives. In 1952 the Canadians John Callaghan and Jack Hopps returned to the idea of the catheter: the electrode is introduced through a vein and the source of electricity is a **battery** attached to the body. The idea was slow to be accepted, because

of the short life of the batteries. It gained acceptance in 1957 when the American Carl Lillehei made **long-life mercury batteries** and reduced the size of the stimulator, which was carried on a belt.

In 1956 however, using the new **silicon transistors** (see p. 84), the American Wilson Greatbach made the first cardiac stimulator, or **pacemaker**, which was miniaturized and, above all, could be implanted. The first implantation took place on a man named Arne Linsson on 8 November 1958, at the Karolinska Institute in Stockholm. Since then pacemakers equipped with **plutonium-238 generators** have been proposed, but it was **lithium batteries** which dominated the market during the 1980s.

The most recent pacemakers record the patient's electrocardiogram and adjust their action accordingly.

Cardiac surgery

Gross, 1948; Hufnagel, 1952; Gibbon, 1953; Barnard, 1967; Cooley, 1969

Surgical heart operations are on the border between development and actual invention, and so they are more appropriately defined as innovations. The first recorded heart operation was the repair of a cardiac lesion in 1896 by the German Ludwig Rehn. For many years it was considered to be an unprecedented audacity, which neither surgeons nor the patients were eager to repeat. Nevertheless the advent of **endotracheal intubation** using **pressurized anaesthetic gas** (direct administration of an anaesthetic via the larynx) in 1904 meant at least that operations inside the thorax were possible, in reasonably safe conditions and without disturbing respiratory function. Some highly skilled surgeons applied themselves to experimentation on the repair of large arteries and in this respect the year 1948 is considered to be a major date: it was then that the American Robert Gross had the idea of suturing an open aorta (large vessel leading from the heart) using artificial elements and tissues

grafted from the patient himself (**autograft**); the operation was crowned with success.

In 1923 attempts had even been made to repair the valves inside the heart. For lack of a really satisfactory technique the results were indecisive. During the 1950s there were plans to graft **artificial valves**; it was the American C. Hufnagel who achieved a successful implantation in 1952 (subsequently the technique was to be considerably improved, as a result in particular of the making of materials compatible with the blood environment).

The following year the American John Gibbon carried out the first **open heart operation**, with the help of a machine which was connected on one side to the large vessels and on the other to the windpipe. This ensured oxygenation and the circulation of the patient's blood during the operation; with the help of his wife and his colleagues he had taken 19 years to complete this **extracorporeal circulation** (i.e. blood flows outside the body to enable oxygenation) apparatus, and the title of inventor belongs undeniably to him. From this time onwards, considerable progress was possible in cardiac surgery which enabled reliable operatory protocols to be established. It was as a result of this invention that in 1967 the South African Christiaan Barnard managed (after numerous attempts on animals) the first **heart transplant**.

Cocaine
Niemann, 1859

Cocaine was extracted for the first time in 1859 by the German Albert Niemann, from leaves of *Erythroxylon coca*, a small shrub known since the destruction of the Inca Empire by the Spanish conquistador Pizarro in 1532. Until then the plantations had belonged to the Inca overlords and great priests. The stimulant properties of the leaves had also been known for a long time. Cocaine was used as an **anaesthetic** for the first time in 1884 by the Austrian Carl Koller.

Cochlear implant
Von Bekesy, 1960; 3 M, 1973

A cochlear implant is a **miniature prosthesis** which — at the end of the 1980s only partially — restores sound for the profoundly deaf. The prosthesis consists of flat components which are implanted under the skin just above the ear and act as antennae to pick up external sounds. These antennae are connected to a receiver which is extended by a special electrical wire, or electrode, which is slipped inside the **cochlea** (inner ear). The sound waves received externally are thus transformed into electric impulses which go to stimulate the auditory nerve linked to the cochlea and thence are transmitted to the brain. In the most recent models, made in 1988 by the American firm 3M and by the Australian firm Nucleus, the sounds were broken down according to their **frequencies** (in the Australian model the breakdown was effected by the use of multiple electrodes, each transmitting a given frequency).

It is not certain who deserves the credit for the cochlear implant; it seems that this type of prosthesis is a development of the kind which were implanted in 1952 in people suffering from advanced otosclerosis, but who went on to have the **stirrup** (an ear bone) rebuilt using prosthetics (stapedectomy). It was certainly this type of operation, which would have been impossible previously, which paved the way to the first experiments with cochlear implants. In 1960 the first to publish his theory was the Hungarian-American, Georg von Bekesy. The first implant was made in 1973 after much theoretical and practical study, by the above-mentioned firm, 3M. The cochlear implant does not restore normal hearing; it simply enables the deaf person to perceive external sounds, and to distinguish them as **low-pitched** or **shrill**. With the help of lip-reading they can re-establish their auditory contact with the human environment.

In 1988 the cochlear implant was only at its primary stage; it was hoped that it would be possible to improve the sound reception sufficiently to allow the restoration of nearly normal hearing eventually.

Contact lenses
Fick, 1887; I.G. Farben, 1936; Softsite CLL, 1985

It was the German ophthalmologist A. E. Fick who was the first in 1887 to devise and make **corneal lenses**, now known as contact lenses. These prostheses were then made of glass and were quite rare. In 1936 the German firm I.G. Farben took up Fick's idea and made some lenses out of **plexiglass** which were much lighter though their manufacture and use still posed too many problems for them to come into general use. Progress was made in 1956 by the Englishman Norman Bier who made some lenses of a much smaller diameter which did not cover too large a section of the eyeball; they were made of **methacrylate** and so were easier to insert and take out.

The most recent improvement is due to the American firm Softsite Contact Lens Laboratory, who make **soft** contact lenses which are even kinder on the eye.

> There is not yet available any type of contact lens which has won the unanimous support of ophthalmologists and can be worn continuously without the use of a hydrating fluid.

Contraceptives
Pincus, 1950

Contraceptives are commonly thought as being an invention and, moreover, a modern invention. This is not the case at all and they are only mentioned here as a matter of interest.

Human beings have practised contraception since very early times. Records as varied as they are old have been found, for example in the Petri Egyptian papyrus, dated around 1850 BC and the Ebers papyrus, dated around 1500 BC concerning instruments of both control and (at the same time) infanticide. It is likely that the methods indicated in these documents had even more ancient origins, such as the use of **vaginal tampons** impregnated with various substances such as honey, powdered crocodile faeces, olive oil, onion juice and mint sap.

The condom (named, it is said, after an Englishman called Condom, who was a contemporary of Charles II and who must have effected its manufacture) appeared in the 16th century, as an accessory for the prevention of **venereal diseases**. It was the Italian doctor Gabriele Fallopio who thought of using segments of animal intestines, 'cleaned' beforehand, and knotted at one end. From 1840 they were made from **vulcanized rubber**, which meant some reservations regarding their comfort. They have been made from latex only since 1930.

As for intravaginal contraception, since the time of the Greeks and the Romans an impressive catalogue of substances with which the tampons had to be impregnated has been used: oil, vinegar, alum, hemlock, green tea, zinc, lead sulphate, quinine, tanin, opium, prussic acid, iodine, strychnine, alcohol . . ., without counting the recipes based on plant extracts which defies inventory. Diaphragms, which are derived from vaginal tampons, are not really a modern invention either: in the 10th century the Japanese used discs of bamboo paper. **Intra-uterine devices** (IUD or coil) appeared in the 19th century and, though they bear no name of particular inventors, they perhaps merit the label of invention more. They consist of a simplification of the tampon, which can be applied to the neck of the womb, the most unexpected of which is undoubtedly the scapular bone of the chicken . . . The first scientific coil was actually made by the German doctor R. Richter.

The contraceptive pill is no more an invention: it proceeds from two discoveries, one by the Austrian Ludwig Haberlandt, in 1921, which was to do with human **anti-ovulatory** secretions, the other by the Americans Allen and Corner in 1929, which was the hormone **progesterone**. In 1941 the American Russell Marker discovered a plant which has been used as a contraceptive by the Mexicans since time immemorial, *Lithospermum ruderale*, which contains progesterone. It was the American Gregor Pincus at the end of the 1950s who gave scientific approval for the use of progesterone for contraceptive purposes by determining the correct amounts to use. Thus the 'pill' was born.

Decompression chamber

Haldane, 1907; Davis and Siebe-Gorman, 1929

It had been observed over several centuries that when **underwater divers** surfaced they were subject to physiological injuries which became more frequent and serious as the dives became deeper. These injuries, which were sometimes fatal, were due to the sharp changes in the concentration levels of the inhaled gases in the blood and the tissues as the divers surfaced. In certain cases — as was later known — this caused the formation of bubbles of gas in the blood circulatory system. The solution to this danger came from the Briton Sir Robert H. Davis from the firm Siebe-Gorman Ltd., who had the idea of putting the divers when they surfaced through **stages of compression** in a decompression box or chamber where the pressure was progressively but slowly lowered. This followed the studies of John Scott Haldane in 1907.

Electrocardiograph (ECG)

Einthoven, 1903

The first electrocardiograph (or ECG) was a **wire galvanometer** invented by the Dutchman Willem Einthoven for measuring the **differences in electrical potential** generated by **cardiac contractions**. Einthoven, who coined the terms electrocardiograph and electrocardiogram, spent the next five years working out the nature of diseased hearts, in order to establish cardiac pathology.

Electronic eye

Brindley, 1968; Dobelle, c.1972

One of the most daring ideas in **prosthetics** was made in 1968 by the Briton Giles Brindley, from Cambridge University; it was an electronic prosthesis of the human eye, designed for blind people. Brindley and his colleagues had worked from the fact that when the **visual cortex** (the part of the brain that controls sight) of a blind person is electrically stimulated using electrodes, he or she sees luminous spots or **phosphenes**. By activating several electrodes at once Brindley released predictable combinations of phosphenes in his subjects, which more or less correspond to the order of the activated electrodes. He went on to devise a system of glasses in which the images would stimulate photoelectric cells which in turn would stimulate some corresponding electrodes fixed to the skull of the blind person.

Around the year 1972 the American William H. Dobelle from Utah University took up Brindley's ideas and research again, with a view to constructing a complete system of image reproduction by 'phosphene points'. This could be comparable to a television image but, considering its elementary nature, it should rather be likened to a notice board of illuminated news reels or the scoreboard in a sports stadium. In *Electronics* magazine of 24 January 1974 Dobelle published the outline of his invention, the cost of which he estimated to be some $5000. It consisted of a false pair of glasses (see diagram) comprising a **miniature television camera**. This would be connected to the visual cortex by a battery of electrodes which could be attached using a Teflon plate. In 1980 Dobelle, using 64-electrode batteries, obtained noteworthy results in six patients: they could identify some simple geometric shapes and even a few letters of the alphabet. The demonstration took place in front of the College of Medicine and Surgery of Columbia University in New York. To this day Dobelle's invention has not become the object of commercial research.

According to the American Richard Normann, of Utah University, the reason for this shortcoming would be the weak resolution power of the apparatus. Normann, who had followed the research, thought the **intracranial implantation** of electrodes was required (which differed from Dobelle's method which applied the electrodes only to the scalp). For Normann the electrodes had to be implanted in such a way that they stimulated the visual cortex directly, penetrating to 1.5 mm below the surface of the brain — which allowed electric intensities of only a tenth to a hundredth of that used in Dobelle's instrument to be used. Moreover, this implantation would avoid the unpredictable dispersal of the phosphenes, which was inherent in the superficial electrodes.

At the end of the 1980s another team, that of Terry Hambrecht from the National Institutes of Health in Bethesda, also researched a type of implantable prosthesis, based on the use of only 20 electrodes instead of the 64 advocated by Normann.

Taking this invention to the practical stage obviously poses some significant problems, given that it is a prosthesis involving recourse to **neurological microsurgery**, with permanent openings in the skull, which present obvious risks of infection. The invention will undoubtedly be a decisive step forward when electrodes which can be stimulated using an **electromagnetic field** are made.

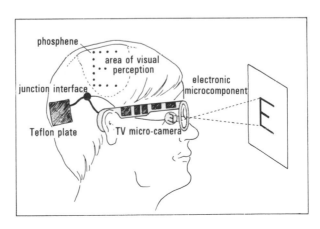

The electronic eye consists of spectacles equipped with two miniature video cameras. The optical signals are transmitted by a circuit connected to an electronic implant: the brain perceives images which are undoubtedly blurred, but which would enable a blind person to get his bearings in a reasonably reliable way.

Galenic medicines

Ehrlich, c.1950

It is almost impossible to assign either a precise date or a person to the invention of 'galenic' medicines (named after Galen, a Greek physician), which constitute one of the most important inventions of the 20th century and of **pharmacology** in particular.

The label 'galenic' refers to a medicine which is ready to be taken, having been prepared by the pharmacist; it is an unsuitable term but has nevertheless been successful, just like the rather unscientific expression 'magic bullets' which is often connected with it. The more precise definition to date would be that of **punctual medicines**.

This 'third generation' pharmacology (the first having been that of the real galenic medicines, prepared to order, and the second, that of specialities prepared industrially) includes a certain number of product types which all have the following point in common: they do not affect the whole organism like their predecessors did; indeed, as they proceed along their customary journey, be it by the **circulatory system** or the **digestive system**, ordinary medicines affect the healthy tissues which they cross over just as much as the target cells at which they are aimed. Hence the **side effects**, called **iatrogens** which follow and which cause their administration sometimes to be toxic. Moreover, they often forbid the use of **crossed medicines** which either make certain medicines possible or else inhibit their use.

It is in this way that tranquillizers affect the intestinal motor functions. Punctual medicines on the other hand, which can be described as 'intelligent projectiles', cross over the healthy tissues without affecting them and attach themselves only to the target cells which have **receptors**.

It seems to have been the American John Ehrlich, from the firm Parke Davis, who during the 1950s was the first to think of medicines with special affinities for certain tissues. They would necessitate a significantly smaller dosage than the classical medicines, of which a more or less appreciable part was destroyed whilst in transit in the blood and the digestion. But the pharmacology of the 1950s was not yet ready to manufacture such products, since **molecular biology** had not reached the degree of perfection which it has today.

It was only at the beginning of the 1980s that the medicine market began to tail off (due to the limited duration of the exploitation of a patent) and that research in this domain recommenced. The specific inventions nonetheless get lost in the crossed networks of the variously different patents, and it is therefore extremely difficult, if not unwise, to attribute the merits of the invention of such-and-such a type of punctual medicine to such-and-such a laboratory. The fact remains that at the end of the decade in question four large types could be differentiated.

The **transdermics** come in the form of **adhesive tablets** which are applied to the skin at determined points of the body. In 1988, by way of example, the American firm Hercon made some transdermics consisting of **non-porous polymers**, arranged in 'mille-feuille', which could carry up to their weight in active molecules. The advantage of this formula lies in the weight of the molecules, which is less than 500 atomic mass units (a.m.u. where one a.m.u. is equal to the mass of a proton), which had never been available in pharmacology. Indeed, above 500 a.m.u. the skin constitutes a barrier which allows nothing to pass through. Moreover, the same firm was researching chemical compounds which could open the dermal barrier to allow the diffusion of various substances, such as protein molecules. Also included in trans-

One of the paramedical areas which had adopted microparticles at the end of 1980s was **cosmetology**. A whole range of new products was in the midst of being prepared, among which were creams capable of effectively releasing substances into the dermis (inner layers of skin) which, despite the promises of the advertisements, could not then cross the dermal barrier, which consists of 20 layers of cells.

dermics is the powder which was devised and made by the American firm Advanced Polymers, which disappears as soon as it is applied to the skin. This 'powder' consists of infinitesimal micro-sponges of 5–300 microns in diameter, each weighing 1/120 millionth of a gram. Once these micro-sponges, which resemble chemical traps like the **cryptans** (see p. 27) have been absorbed by the skin, the substances with which they have been charged are distributed by the alteration of the environment in which they find themselves.

Osmotics are very small molecules which are injected through the skin without the use of a mechanical instrument, but by **electro-osmosis**; the application of a weak electric current modifies the permeability of the living cells and then enables them to absorb molecules which would not cross them in ordinary conditions.

In 1988 the American firm Drug Delivery Systems Inc. succeeded in obtaining, in an animal, concentrations of insulin equal to those of ordinary subcutaneous injections using a needle. This had been achieved using Powerpatch adhesive tablets, and by applying sufficient electric current to the epidermis (skin surface).

Molecule reservoirs obey a principle which is roughly comparable to that of the **depot medicines**, from which the molecules only diffuse slowly in a given environment, with the slight difference that in their new form they can cross certain tissues which were previously thought of as unsurmountable barriers.

Several American (Alza, Elan, K.V. Pharm, Penwalt, Verex Labs) and European (INSERM, France; Sopar, Belgium; Nordisk, Denmark) firms in 1988 made microcapsule-reservoirs with a diameter of 250 nanometres out of biodegradable polymer. These were harmless and could pass through the intestinal mucus without being dissolved; the release of the product — in this case insulin made to the formula of the INSERM   only occurs at the sub-mucus level. This already constitutes a fundamentally new method of treating diabetes. **Osmotic micro-pumps** of the Oros model also enter into the molecule-reservoir group, the principle of which was patented by Alza. These really are microscopic pumps which function by **osmosis**, that is by absorption of the intestinal liquid, and which diffuse medicines continuously into the intestine (hormones, anti-inflammatories, relaxants, vitamins, etc.). Their advantage is in that the concentration does not decrease between two oral or injected doses of a medicine, so the number of dosages is reduced and some treatments are made much more convenient.

Finally, **aerosol galenics** make profitable use of the high absorption capacity of **nasal mucus** and progress made in the micro-miniaturization of molecules in order to diffuse products as diverse as hormones, insulin, anti-inflammatories, etc. into the blood.

One of the most promising prospects of punctual pharmacology at the end of the 1980s was the expansion of **polymer** micromolecules capable of crossing the meninges (lining of the brain), which was also reputed to be insurmountable. They are diffused in the bloodstream and can thus reach the brain to release various products such as tranquillizers and anti-cancer drugs.

At the same time **liposomes**, or micromolecules of fat, were also a very active area of study. Although they were largely conclusive, the first studies nevertheless revealed some immunological problems: the liposomes are more easily recognizable by the immune system, and so are 'swallowed' by the macrophages before having released their contents.

Another prospect being explored was the direct administration of **corrector genes** for certain metabolic defects which bears an even more considerable significance than the preceding products.

Gamma helmet

Leksell and Larsson, 1968

The gamma helmet, which is also known under the name **gamma knife** is an accessory of **radiotherapy**. When fixed to the skull, the helmet enables the irradiation with **cobalt-60** to a precision of a hundredth of a centimetre for the treatment of **brain tumours**. The source of the cobalt-60 is incorporated into the helmet itself and produces radioactive beams following the stereotactic (i.e. three-dimensional) location of the tumour. The instrument was invented in 1968 by the Swedes Lars Leksell and Borje Larsson. Although with 500 patients its use had a success rate of 87% in 1988, because of its cost, there were only five such instruments in the world. The manufacturer is Elekta Instruments of Sweden.

Genetic grafting

Boyer, 1977

One of the basic operations in **molecular biology** is the genetic graft; this consists of grafting a foreign fragment of the **genotype** or **DNA** (for **deoxyribonucleic acid**) of an organism into another organism. This operation was carried out for the first time by the American Herbert W. Boyer, who fused a segment of human DNA which had been synthesized in a laboratory, to a bacterium *Escherichia coli*. The segment in question was that of a cerebral (brain) hormone, **somatostatin**, which controls growth. The bacterium began to produce somatostatin.

This operation had, however, been preceded in 1974 by the insertion of genes coding the **resistance of bacteria to penicillin**, extracted from penicillin-resistant bacteria **plasmids** (pieces of genetic material) into a plasmid of *E. coli* which did not have this resistance. By extending this it could be said that this operation, which was carried out by the American Stanley Cohen's research group at Stanford in California, constitutes the precursory stage to genetic grafts, although it does not belong in the specific area of bacterial genotype; indeed, a plasmid is an organ of the bacterium which does transmit genetic information, but it does not constitute the essential part of the bacterial DNA, which is found in the **nucleus** of the bacterium (whereas the plasmid is found in the **protoplasm** surrounding the nucleus). Boyer's invention was a feat which did not have immediate industrial repercussions, for the precise properties of somatostatin were still not known; however it opened the road to genetic grafting since extended to the treatment of genetic diseases.

Iron lung
Drinker, 1927

An iron lung or **respirator** is an instrument used for **resuscitation**. The patient's body is placed inside and by assuring **compression** and **decompression** outside the organism it causes movements of the thorax which correspond to the movement of respiration. It was invented in 1927 by the American Philip Drinker of Harvard University and was manufactured by the American firm Warren E. Collins Inc. It has proved invaluable in cases where the respiratory system of the patient was malfunctioning, for example following paralysis.

Lithotripter
Eisenberger and Chaussy, 1972

The fragmentation of kidney stones without haemorrhage using **ultrasonic waves**, was invented in 1972 by the Germans Eisenberger and Chaussy in Munich. A series of shock waves focused on the patient, who is suspended in a special bath, reduce the stones to a powder which is then eliminated in the urine. The lithotripter prototype was commercialized in 1980 by the German firm Dornier GmbH.

Mesotherapy
Pistor, 1952

Mesotherapy, which can be defined as an **intradermotherapy**, is a way of administering medicines by injection into the **mesodermis**, the middle layer of the skin, using a short needle or an apparatus comprising several short needles, called a **multi-injecter.**

Mesotherapy was invented in 1952 by the French doctor Michel Pistor; it is important for two essential reasons, the first being the **reduction in dosage** of injected medicine, the second, a more pronounced **localized action** than that which is obtained by the normal method of drug administration (intravenous or oral). For example, mesotherapy of otitis (ear inflammation) can be treated by injecting in areas very near to the ear.

Mesotherapy is used in an optional way for the treatment of pains and by some practitioners it is associated with a treatment based on **procaine** or **lidocaine** (local anaesthetics) which are injected together with other products.

> The general and local effects of the injection of procaine into the mesodermis have been noticed by many practitioners, including René Leriche (1929). It seems in particular that procaine delays the action of injected medicines and thus prolongs their effect, over and above its **vasodilatory** (expansion of blood vessels), **antihistamine** (anti-allergic) and **relaxing action.**

Monoclonal antibodies
Kohler and Milstein, 1975

Monoclonal antibodies are substances which are identical to the ordinary antibodies secreted by the **immune system** against **antigens**, or agents which are foreign to the organism and therefore potentially dangerous. Their characteristic is that they can be artificially produced by the fusion of two different types of cells: firstly, **cancer cells**, chosen because they have the property of continuous reproduction, unlike healthy cells which stop dividing after 52 or 53 generations, and secondly, **lymphocytes**, white antibody-producing globules. The fusion between a cancer cell and a lymphocyte produces something known in biology as a **chimera**, or a **hybridoma**. This has the advantage of being able to be cultivated indefinitely *in vitro* (in the test tube), which is not possible for the ordinary lymphocyte. A hybridoma was first developed in 1975 by the Britons George Kohler and Cesar Milstein; the discovery was made possible by the choice of a substance which dissolved the membranes of the two cells and thus facilitated their fusion into a single viable cell. It is really a double discovery, for the lymphocyte, when tested on an animal which had been injected with a given antigen, gained the property of being able to make the corresponding antibody. The cultivation of hybridomas thus enabled a theoretically infinite quantity of these antibodies to be collected. The adjective 'monoclonal' is derived from the fact that it is a **clone** (genetic copy) with unique properties.

Several tens of monoclonal antibodies have been obtained since Kohler and Milstein's work. They are used in two ways. The first consists of the diagnosis of organic states, such as pregnancy, metabolic abnormalities, and illnesses by incorporating the monoclonal antibodies into chemical reagents; when in the presence of the specific antigen for which it is made, the antibody will attach itself to it and so indicate its presence. The other use is curative, as the monoclonal antibody reinforces the organism's natural immune defence system. Thus in France in 1987, the monoclonal antibody anti-LFA-1 was used to treat seven children who had a rare bone disease (**osteoporosis**). Subsequently the treatment of cancer is predicted as well as warnings of transplant rejections.

Obstetric ultrasonography
Donald, 1979

The first to have the idea of using ultrasonography to examine the **foetus** and then to do so was the Briton Ian Donald, professor of obstetrics at Glasgow University. This technique turned out to be of great importance because it put an end to obstetric **radiography**, the risks of which had been denounced with increasing fervour since the end of the 1960s, but which had nevertheless constituted the only means of observing the foetus, and in particular its position, before birth.

A few reservations have been expressed as to whether obstetric ultrasonography is really safe. Some checks have led to the opinion that harmful consequences have not been recorded in children who underwent this examination at the foetal stage. At the end of the 1980s it was thought that three examinations carried out during the monitoring of a pregnancy presented no risk at all.

Pernicious anaemia (treatment of)

Folkers and Shorb, 1947

This type of anaemia was traditionally called 'pernicious' because it was eventually a fatal disease, with no specific treatment known. In 1947 Karl August Folkers and Mary Shorb, working in the United States, were the first to use injected **vitamin B$_{12}$** to treat it. The illness is actually the result of an incapacity to absorb this vitamin, due to a deficiency in an element of the gastric mucus, known as an 'intrinsic factor'. It is also known under various other names including: Addison's anaemia, Biermer's anaemia, Runeberg's anaemia and macrocytic achylic anaemic.

Phenotype (and genotype)

Johannsen, 1902

The phenotype, which was devised by the Dane Wilhelm Johannsen in 1902, must be cited as a concept which acts as a tool enabling the clarification of analysis and the guidance of research. This is the **visible expression of genes** in a living organism. At the same time Johannsen, a botanist and geneticist, invented the opposite concept (the genotype), which may or may not be shown, and the expression of which is not clearly visible. Subsequently Johannsen proposed the masterful idea in biology that the same genotype can be expressed in different ways according to its environment.

Psychoanalysis

Breuer and Freud, 1890

Psychoanalysis presents itself as a **technique of treating mental disorders** using the sole expedient of **discourse**, and without recourse to organic, chemical, physical or other means. It is derived from the particular observation of patients under **hypnosis** made by the Viennese physicists Joseph Breuer and Sigmund Freud: the patients freely expressed feelings which they were repressing in their conscious state. These sessions brought about some improvement in their psychological troubles.

Breuer and Freud developed a technique which consisted of encouraging the patients to abandon themselves to the **free association of ideas** without any discursive link, in order to recreate a state analogous to that of hypnosis where the consciousness was asleep. In general the patients had no difficulty in applying themselves to this discipline, but it was Freud who noticed that some of them were repressing the expression of painful memories, or else they did not bring them to consciousness, that is, to the level of discourse. As most of the repressed memories brought back unpleasant sexual experiences, Freud concluded that the **anxiety** from which his patients suffered was caused by the **inhibition** of the sexual instinct, or **libido**. From this premise Freud and his followers extended the notion of anxiety to include phenomena such as aggression and phobias, and they formulated an interpretation of the personality.

According to Freud the personality could be represented by a three-level structure. At the deepest level there was the **id**, a store of genetic determinations and creative force responsible for growth and evolution of the individual, as well as of memories; above this there was the **Ego**, operating in the conscience and carrying out cognitive and perceptive functions and voluntary acts; on top of this was the **Super-ego**, a reservoir of ideas instilled during education by the family and society. It was conflict between the structures which caused abnormal behaviour, hence the anxiety.

Freud also introduced a new notion which was the **transfer** of the love or hate which the patient bore towards his parents to the practitioner, or the psychoanalyst, who was treating him.

Freud's work is best known among the general public by his concept of the subconscious as a domain of urges and memories which corresponds more or less exactly to the id, and by the '**Oedipus complex**', that is, the potentially sexual attachment of any person in his or her infancy to the parent of the opposite sex. (This refers to the legend of Oedipus who unwittingly married his mother Jocasta.)

Psychoanalysis is a discipline where numerous streams of interpretation, often straying from Freud's original doctrine, have created other schools. It was in this way that the Swiss Carl Jung's emphasis on the role of the **collective unconscious** led to the Jungian school, and the German Alfred Adler's emphasis on the role of **social pressures** also led to Adlerian psychoanalysis. In France between the 1930s and the 1980s Jacques Lacan created his own school and introduced concepts such as that of the **mirror stage**. This postulates that a successful personality formation depends on the subject passing through a stage in infancy where he at first does not distinguish between the Ego and the external world, which he considers as a mirror, and then learns to differentiate himself and thus construct his Ego. Lacan also postulated that 'the subconscious is structured like a language' and that the dynamic of the subconscious can only be organized on an axis, that of the desire of the Other.

Between 1920 and 1960 psychoanalysis became widespread in the Western world — but not in Eastern Europe or Asia — as an independent psychological discipline. It was sometimes associated with psychiatry and in some countries could be practised without any medical qualifications. Its cultural significance was considerable. Since the 1960s however, it seems that its expansion has stabilized; it has even declined in the United States where during the Second World War psychoanalysis had found unequivocal support.

It really is an invention, the true value of which has been in defining the domain of the subconscious, which until then was either not known, was neglected or studied little by classical psychologists. Its fault has been in the systematical isolation of physiological foundations from the nervous system, which has resulted in the Freudian theory unfortunately being encroached upon by the occult and by outrageous blunders, such as the attribution of a sexual role to the nose, following the theories of one of Freud's doctor friends, Wilhelm Fliess.

Generally speaking, psychoanalysis has not benefited from the immense progress made in **neurophysiology** which would have enabled it to establish its postulations with more reliability. The discovery of the division of the central nervous system into a **palaeo-cortex** ('ancient brain') and a **neo-cortex** ('new brain') enabled the science to go beyond Freud's distinctions, which were not founded on phenomenology nor on a structural knowledge of the generative neurons of the nervous and intellectual systems. The former is a primitive structure developed since the reptilian stage of evolution of species and is basically the seat of the five primary impulses (hunger, thirst, fear, genetic instinct and aggressiveness); the last is of more recent evolution, and constitutes the physiological place for the rational integration of perceptions and voluntary acts. Research on the **dream**, to the interpretation of which Freud and the psychoanalysts attached such great importance, has revealed moreover that the stimulation of memories in the dream takes place in a semi-vague way, and not following the strict associative logic advocated by the psychoanalysts. Finally, certain uncertainties in the Freudian theory must be mentioned, such as the difficulty of establishing a plausible border between the conscious and the subconscious, and some rather improper interpretations.

The excessive emphasis put on sexuality explains why Freudianism, and other forms of psychoanalysis, has never appealed to Asiatic cultures where sexuality does not in any way bear the taboo character that it has in Judao-Christian cultures. Moreover, **anthropology** and **ethnology** have led to the theory of the Oedipus complex being put into question, for it is unknown in primitive social structures, as research, particularly by Bronislaw Malinowski, has shown. This presumed complex is not fundamental but cultural and geographically limited.

Some excessive interpretations of Freudianism have led ultimately to the wrong conclusion that the pain of anxiety is always a bad thing; it has made France, for example, into the largest consumer of **tranquillizers** in the world; sometimes stress is a beneficial reaction. Moreover, the success of 'popular' Freudianism in the 1960s resulted in the opinion in some circles that any hindrance to the sexuality was generative of anxiety and therefore bad, so the answer was to practise unrestrained sexuality as this was proof of good health. Such exaggerations culminated in results which were the opposite of those which had been naively anticipated, for they led to a **social disintegration** of an often acute nature.

Radiography

Röntgen, 1895

The first **radioscopic image** dates back to 22 December 1895; it was an image of the hand of the wife of the German Wilhelm Konrad von Röntgen who discovered X-rays. It is now known that this type of image is obtained as a result of the intensity difference of the **X-rays** which reach the photographic plate, since they penetrate soft

materials more easily than hard materials such as bone. The significance of Röntgen's invention — the radiographic image, not the X-rays — was universally understood after the publication of the author's paper. Radiography was used extensively during World War I to locate foreign bodies in the flesh of the wounded and for defining fractures. The use of contrast products such as **barium sulphate** began in 1897 or 1898 for the study of hollow organs such as the stomach.

> The first mobile radiology unit was put into use by Marie Curie during the 1914–18 war, so that the injuries of the wounded soldiers could be studied immediately when they were brought behind the battle lines. Marie Curie worked as an operator of one of these units which was transported by motor vehicle.

> The first **radiographic diagnosis** of pulmonary tuberculosis was carried out by the Frenchman Antoine Béclère in 1896.

Rapid blood analyser

Diamond Sensor Systems, 1987

In numerous surgical operations, such as **open heart surgery**, it is necessary to be able to know the level of different constituents of the blood very quickly, to ensure the safety of the patient. In order to obtain this information surgeons have to take blood samples during the operation and send them to the laboratory which analyses them as quickly as possible. In 1987 the American firm Diamond Sensor Systems, from Ann Arbor, Michigan (USA), made and patented a unit of hospital equipment which, when installed in the operating theatre itself, could carry out the analyses automatically in two minutes. When the apparatus, called GEM 6, analysed a 2 cc blood sample it provided a print-out of the state of six essential blood parameters, **oxygen, carbon dioxide, pH (acidity and alkalinity), potassium, calcium** (the **electrolytes**) and **haematocrit** (a measure of blood-cell volume).

Scanner

Hounsfield, 1973

The scanner, also called **scanograph, tomodensitometer** or **computerized tomograph** is an important development in **radiography**. It was invented in 1973 by the Briton Godfrey N. Hounsfield. It works by radiographing part of the anatomy no longer as a whole but in slices of determined thickness. Then, using a computer, it reconstitutes the image according to the **density of the X-rays** absorbed by each slice of the anatomical part, with this density being measured in pixels. Since the normal absorption densities of bone, soft tissues and liquids are known, any abnormal density can be detected and the nature of it determined as, for example, liquid cyst, fatty lipoma, internal bleeding, calcium deposit or tumour. This principle is obviously of considerable importance for **diagnoses** which are much more refined than with classic radiography.

The scanner's originality lies in its system of receiving the very narrow (or collimated) beam of X-rays, consisting of a **scintillator** (a substance that emits a flash of light when struck by X-rays) and a **photomultiplier** (to amplify the signal). It has the additional advantage of exposing the body to the smallest visible amount of X-rays because it uses ultrasensitive film, the impression of which is the same as the appearance of the image on the screen of a cathode tube.

Scanography was slow and inconvenient at first due to the required time of immobilization (from 20 to 80 seconds per slice) but it has become much quicker since the number of **detectors** has been increased.

The choice of the term 'scanner' is due to the fact that the X-ray source actually 'sweeps' or 'scans' by rotation through a given angle.

Serotherapy
Von Behring and Kitasato, 1890

At the end of the 18th century, **immunization by vaccination** was beginning to be practised on a large scale (against smallpox). One century later the principle behind it was understood: when the immune system of an organism comes upon a germ in a weakened state, it produces **antibodies** which can then protect it even against the strongest form of the germ. In 1890 the German Emil von Behring had the idea of treating an infected person by the direct administration of a **serum** extracted from the blood of another infected person, which therefore already contained antibodies.

He improved this idea with the Japanese Shibasaburo Kitasato, and it was put into practice for the first time with the anti-tetanus serum. This idea was the foundation of serotherapy and it has since been extended to many other infections. Serotherapy does not have the lasting effect of **vaccination**, but it has the advantage of being able to be used for curative as well as preventative means, against diphtheria and tetanus, for neutralizing toxins such as those in botulism, or for alleviating certain infections which can lead to complications, such as measles.

Tampon
Hass, 1930

In order to offset the inconveniences of impractical sanitary towels, the tampon was invented in 1930 by the American Earl Hass. It was such a success that in one year it had made the flourishing firm founded by Hass, the Tampax Co. Since then the formula has been improved many times; modifications were made following evidence of side effects of infection by the **anaerobic** (without oxygen) **proliferation of certain germs**.

Thermography
Haxthausen, 1932; Lawson, 1957; Leroy, 1980

Thermography is a technique of examining the human body and all objects in general which emit infra-red radiation due to their body heat. The photographically recorded differences in intensity provide information on the thermal fluxes which are invisible to the naked eye.

The 'father' of thermography is thought to be the famous English astronomer Sir William Herschel who was the first to study the heating effects of infra-red. In 1932 the Dane O. Haxthausen thought of applying the principle of infra-red radiation from the human body to that which he called **subcutaneous photography**. He must therefore be regarded as the inventor of this medical technique, which could not be brought into general use due to the slow development of sensitive films at the time. Nevertheless it was used in **aerial photography** during World War II to detect things which could not be seen in visible light, such as camouflaged factories, whose infra-red radiation showed that they were in use, as well as concentrations of military vehicles. This technique was to lead to a discipline of agricultural and geological study by aerial reconnaissance known not under the name of thermography, but of **infra-red photography**. The difference between thermography and infra-red photography is that the latter does not 'photograph by heat' as such, but rather by the reflection or absorption of infra-red radiation. Therefore the green paint for the camouflage of a truck or a factory absorbs infra-red whilst the green of foliage reflects it. The green of the truck or factory appears in blue or yellow on the photograph, where-as the foliage will be red. After the war the improved quality of photographic film allowed Haxthausen's principle to be taken up again and useful negatives of the human body to be obtained; the first to use this technique was the American John Llewellyn Lawson in 1957.

Thermography is particularly useful in research on **breast tumours**. In 1980 the Frenchman Yves Leroy developed an appreciably finer version, **short-wave thermography**. Thermography is currently used for the thermic surveillance of certain industrial plants and for monitoring the insulation of buildings.

Tranquillizers
Laborit, 1952; Sternbach and Randall, 1960

The first tranquillizer was **chlorproma-zine**, a **phenothiazine** derivative, which is defined specifically as a **neuroleptic** (a drug which reduces nervous tension), and which was discovered and made in 1952 by the Frenchman Henri Laborit. It was first used in that year by Jean Delay and Jean Deniker for calming mentally ill people who were having a fit, because it had two effects: one was anti-hallucinatory and sedative, due to the reaction of the subst-ance with regard to the **dopamine** in the brain; the other, more particularly seda-tive, was due to the reaction with regard to the **noradrenaline**. It was commercialized under the name of Largactil and gave rise to numerous derived molecules. This tran-quillizer was followed by Librium and Valium which were made by the American Lowe Randall on the basis of observations made by the American Leo Sternbach. These had sedative effects from another class of chemicals, the **benzodiazepines**. The specific definition of these is **psycho-trops** (behaviour influencing); they have different effects to chlorpromazine — seda-tion, anxiolysis (reducing anxiety), muscular relaxation and anti-convulsive action.

Transgenic mouse
Leder and Stewart, Weissmann and Mosier, 1988

In March 1988 the American biologists Philip Leder and Timothy Stewart requested and obtained a patent for the creation of a mouse whose **genetic code** had been altered. The alteration consisted of the insertion of the gene *c-myc*, an **oncogene** or gene which makes the organ-ism susceptible to the development of a cancer. This was done with the aim of studying the carcinogenic elements in the environment. When these mice were sub-jected to potential carcinogens they would manifest toxic reactions to these substances. The granting of an invention patent raised a debate in the scientific community because it implies the **industrial owner-**

ship of a living organism, a concept which is as difficult to integrate into jurisprudence as into scientific ethics, additionally so because during the preceding years numerous organisms whose genetic code had been modified, called transgenic organisms, had been created without being patented. In this way, for example, **transgenic bacteria**, which can make insulin, had been made, as well as transgenic cows, capable of producing milk with a different fat content. In 1988 the transgenic organisms numbered several hundred. The existence of a patent for Leder and Stewart's special mice meant that no one had the right to create equivalent ones and that they were the exclusive property of their inventors.

In September 1988 the transgenic mouse was the focus of a second fundamentally different invention. Within the context of the research on a **treatment for AIDS**, particularly by **vaccination**, the biologists came up against two difficulties. The first was the growing sparsity of laboratory monkeys, due to increasingly strict regulations. The monkey was the only animal capable of providing a model, albeit approximate, but nevertheless useful, for the study of the reactions of the human immune system. The second was related to this same problem, given that the monkey does not have exactly the same immune system as man. In a domain as precise as the study of a vaccine it is almost impossible to imagine it being produced commercially without the real effects on the human being known, but humans obviously cannot be used in experimentation.

It was then that the Americans Weissmann and Mosier had the idea of making use of the exceptional characteristics of a certain species of mouse; this was the **SICD mouse** (for Severe Immuno-Combined Deficiency), which is born devoid of any immune system as a result of a hereditary genetic deficiency. These mice have to live in completely aseptic bubbles, otherwise they die. Weissmann and Mosier grafted fragments of **human foetal tissues** onto these mice. The tissue was taken from the liver, bone marrow, spleen, the lymph nodes and the thymus, all principal organs which assure the production of lymphocytes and other cells charged with immunity. The fact that the SICD mice had no immune defences is the reason why they tolerated the transplants without any problem. Then as the foetal grafts developed, they produced antibodies, that is, the animal became equipped with immune defences. The most interesting point is that it developed an identical system to that of man. Therefore the subjects were equivalent to human beings, at least from the point of view of immunity, which was what interested the biologists in the first place. The mice therefore replaced the monkeys and during the winter 1988–9 work was to begin on the anti-AIDS vaccines.

This invention went far beyond the domain of AIDS, for it allowed the experimentation on many other vaccines and **anticancer treatments** in particular; so it received an enthusiastic welcome from the scientific community. Nevertheless the use of foetal tissues posed **ethical** problems which seemed impossible to solve except by the modification of the American conventions on the use of such tissues.

Water with hydrogen and oxygen in isotopic form

Lifson, 1959

Water, with isotopes in the form of **deuterium** (or **heavy water**, D_2O) and **oxygen-18,** is the finest and simplest instrument for evaluating precisely the quantity of energy necessary to daily human activity. Conceived in 1959 by the American Nathan Lifson, its use in the study of the **metabolism** is based on the fact that this is linked to the exchanges of water within and between cells. Knowledge of the rate of water exchange according to the automatically constant body temperature (37°C) ought to enable the energy spent by an organism to be estimated. However, knowledge of these rates remained unattainable, largely because the water in the body, which is called **free water**, could not precisely be distinguished from **cellular** (or bound) **water**. So it was not possible to establish their respective dynamics in relation to time, which constitutes an essential factor of the equation on which metabolism is based. Energy was therefore calculated on the basis of the oxygen consumption and carbon dioxide exhalation by a person at rest, who was not allowed to eat — hardly a very convenient method. On the other hand, urine analysis by **mass spectrometry** carried out on an individual who had absorbed non-degradable and relatively non-toxic chemicals in determined quantities could enable more precise measurement of the rate of exchange according to their temporal excretion curve. This was Lifson's

idea which could be implemented only in the second half of the 1980s as a result of progress made in mass spectrometry. It earned its inventor a scientific prize (the Rankin prize) in 1987.

Attempts to calculate the basal metabolism using this method led to a proportional lowering of the assessment of the basal metabolic rate or BMR, which was from 9 to 25% depending on the case in question (age, sex, genetic predisposition, activity, etc.). It revealed that overeating increased the metabolic rate and offered a basis for the explanation of the activities of certain animals, such as birds, which travel over large distances without feeding. It should be remembered though that the organism is comparable to a black box, where all that is known for certain is what goes in and what comes out; the internal transformations which have taken place are then deduced. Metabolic calculations therefore include a significant margin of uncertainty.

Yellow mercury oxide ointment
Pagenstecher, 1862

The use of the antiseptic properties of a mercurial derivative, yellow mercury oxide, in an ointment for ophthalmological treatment was inaugurated in 1862 by the

German Alexander Pagenstecher. For this reason the salve was first called Pagenstecher ointment. Its use survives into the 1990s.

Transport

The period since 1850 has seen some dazzling transport. The flight of the Wright brothers inaugurated a development which affects not only the area of transport itself but also the whole of world culture. Man was able to cross in a few hours distances which had previously been perceived as immense, and so his world view changed. Astronautics, the sister of aviation, was to change the way man viewed the universe itself. Because of air travel, tourism developed into an industry of major economic importance, but with astronautics, which opens (or half-opens) the doors of infinity, man paradoxically began to experience a sense of his own mortality.

Another development of major importance was the car, the destiny of which has been changing with a disconcerting rapidity ever since it became so universally accessible to the inhabitants of industrial countries. Originally considered as an instrument of freedom, since the 1960s the car has seemed to be more a disposable consumer product; it was thought of next as an economic problem, before being considered by some as a danger to the health of humanity. From a strictly economic point of view, at the end of the 1980s many countries were increasingly aware of the growing cost of road accidents.

Although infinitely more limited, the bathyscaphe and its derivatives, the exploration submarines, deserve a special mention in the history of transport, for they contribute to the growing knowledge of the last unspoilt areas of the Earth, the great ocean depths. Indeed, submarine exploration will have provided more knowledge of the globe than any other means of travel.

Aerobie
Adler, 1985

The object capable of gliding the greatest distance when thrown (100 m) was made in 1985 by the American Alan Adler, a professional inventor. It consists of a flexible ring, the cut shape of which has been worked out by a computer.

Aeroplane
Ader, 1890; Leduc, 1939

Since the beginning of the 20th century there has been endless quibbling over the recognition of the inventor who made the first flight propelled by **mechanical energy**. Precedence goes back incontestably however to the Frenchman Clément Ader (1841–1925) who, on 9 October 1890, succeeded in getting an apparatus heavier than air to take off. It had wings like those of a bat and a propeller powered by steam motor. Ader covered a distance which according to witnesses varies between 50 and 60 m. It was Ader who invented the word 'aeroplane'. His apparatus was called *Eole*.

Ader then built an improved model, *Aeroplane III*, which also had a steam motor, whose performance seems to have been less conclusive. An official report which was mysteriously kept secret until 1910 states that the machine did not fly, whereas Ader affirms that it covered a distance of 300 m. Be that as it may, the following stage which was the propulsion of a machine heavier than air this time by an internal combustion engine, was achieved by the Americans Orville and Wilbur Wright who covered 284 m at a height of several metres.

In fact the Wright brothers benefited from the experience of their predecessors, in particular the following ones: the German Otto Lilienthal who experimented with **gliders** (which led him to the conclusion that the front edge of the wing had to be raised); the Frenchman Robert Esnault-Pelterie who, in the same year 1903, equipped his gliders with ailerons (control surfaces) which could be moved using cords; the Franco-American Octave Chanute who was the first to build gliders with two or three aerofoils, later called **biplanes** and **triplanes**; and the American Samuel Pierpoint Langler who was the first to equip gliders with internal combustion engines, though he did not achieve any conclusive results.

These precursors all met with failure (some tragically, like Lilienthal and Chanute), for many different reasons. These were dominated by the breaking up in the air of the excessively light and pliable structures which they had built, and were no doubt exacerbated by the lack of stability caused by ignorance about the way the ailerons functioned. The Wright brothers adopted Chanute's glider in its biplane version, and the Esnault-Pelterie elevator which they installed at the front. Their inaugural flight was preceded by several flights of about 50 m. The Wright brothers had many successors: Alberto Santos-Dumant, Léon Levasseur, Louis Blériot, Glenn Curtiss, Charles and Gabriel Voisin, Samuel Cody, Alliott Vernon Roe, Igo Etrich . . .

Aviation had been born and was to go through the stages of the **metallic cabins** following those made of wood and canvas of the multimotor aeroplanes, then **aerodynamization**, which, together with the increase in power of the engines would lead eventually to the first **long distance flights** and finally to the creation of the commercial airlines.

Aeroplanes were originally linked in design and performance to the internal combustion engine. The next stage was that of the jet engine. This enabled such a great increase in speed that it quickly caused a revision of the aerodynamic shapes, then of the actual structure of the airframe, and, finally, the materials. This technical evolution was again accelerated with the progress in **electronics** and particularly with the possibility of installing **control circuits** for the handling of the aeroplane in the air. Its commercial consequences were immense, because in more or less halving the time needed for travelling a given distance, the jet enabled the communication network to spread to regions which had not previously been included—far-flung areas of the Pacific, Asia, Africa and South America.

The jet aeroplane had been sketched out briefly in 1928 when the Germans Alexander Lippisch, a glider builder, Fritz von Opel, an engineer, and August Sander, rocket maker, had equipped a glider of the Storch series with a rocket engine called *Ente* ('Duck'). This was not followed up, because the engine would not work for more than quarter of an hour. As for Leduc, he took up the theory set out in 1910 by René Lorin on the possibility of propelling a craft heavier than air by **thermoreaction**, and invented a type of extremely simple jet engine, the **ramjet**, which had no revolving parts: the air entering into the mouth of the engine was compressed there by aerodynamic effect, then was directed towards a combustion chamber which was lined with fuel injectors; the shape of this latter chamber, narrowed at one end, put additional pressure on the hot gases, the ejection of which assured the propulsion. Work on the aeroplane began in 1936 and the prototype was completed in 1939. Its development was interrupted by the war.

In the UK, Sir Frank Whittle (see p. 209), had been attempting to create a turbojet engine between 1928 and 1930. In a turbojet the exhaust gases create the thrust of the engine, driving the machine to which it is attached forward. The Gloster Whittle E28 flew in 1941. Similar work in Germany by Hans Pabst von Ohain culminated in the flight of a Heinkel fighter plane in 1939. In 1945 Rolls-Royce built the first turbojet

with an afterburner. The afterburner allows greater thrust without much increase in engine weight, but it has a very high fuel consumption, so its use tends to be restricted to take-off.

In Italy Secondo Campini had made a turboreactor aeroplane, where the admitted air was compressed at three stages before passing into the combustion chamber; at the exit the thrust of gas was increased by a post-combustion chamber. In 1941 the Caproni-Campini CC2 covered the distance from Milan to Rome at the record speed of 500 km/h. Oddly enough, this experiment had no sequel, at least in the domain of military aircraft construction, the only area in which there was activity at that time, and despite amazing results obtained by two revolutionary craft in Germany. These were firstly the Messerschmidt twin-engined jet ME-163 which was the first successful jet engine in aviation history — and which had been invented by the Alexander Lippisch mentioned above; next and in particular there were the results of the ME-163 VI *Komet*, which on 13 August 1941, piloted by Hans Dittmar, was the first in aviation history to verge on the speed of sound at 1003 km/h (Mach 0.84). Only a small number of the *Komet* were made, a craft with neither empennage (tail unit), nor undercarriage, due to technical problems which arose during construction and especially the risk of leakage of the two fuel tanks of liquid oxygen.

Ader had numerous forerunners who tried to fly **heavier-than-air machines**, including Félix du Temple (1857), the Englishman Thomas Moy (1875), the Russian Alexandre Mozhaisky (1884), the Frenchman M. A. Goupil (1884) and an immediate successor, the Englishman Horatio Philips (1893). Only Mozhaisky, who launched his aircraft from the top of a ski jump and managed to stay aloft for a few seconds, succeeded in a 'take-off', which prompted his fellow countrymen to claim to have achieved the first mechanical flight.

All-metal bodywork

Napier, 1902

When the automobile industry first began, metal was only used for certain parts of the body, those which would withstand distortion the least, particularly the chassis and the doors, as well as the door frames. The floor and the roof were usually made of wood, either with the frame made of wood and covered with waterproofed canvas, or made totally of wood.

The first motor-car builder to have the idea of making a **completely metal body** was the firm Napier in 1902, with its *Napier 9*. Insufficient rigidity and difficulty in repairing the damaged parts caused a delay in the adoption of the all-metal body, which did not come into widespread use until the 1920s.

Aqualung

Cousteau and Gagnan, 1943

The first system which would allow a diver the freedom to swim without being attached to a surface boat was invented and made in 1943 by Jacques-Yves Cousteau and Emile Gagnan. The gas used was **compressed air**. The originality of the invention, to which Gagnan brought his engineer's specialist abilities in the manufacture of control valves, was the **demand valve**,

which supplied the gas to the diver through the mouthpiece at a pressure corresponding to that of the depth at which he was. This invention made possible the exploration of reefs and underwater habitats, both for scientific research and for recreation. The mixture now used is **helyox**, a helium–hydrogen–oxygen mixture.

Astronautics

Tsiolkovski, 1883–1929; Esnault-Pelterie, 1912; Goddard, 1916; Oberth, 1929

Astronautics is derived from the science of rockets which preceded it by several centuries (see *Great Inventions Through History*). The imagination had to take only one step from the use of rockets capable of rising several hundred metres, to expeditions beyond the bounds of the Earth. In 1883, in a paper where he outlined the mathematical and physical principles of rocket motors, the Russian Konstantin Tsiolkovski mentioned the rational possibility of voyages in space and notably the principle of the multi-stage rocket to attain escape velocity.

The development of astronautics depended on mechanical calculations establishing the escape velocity of a projectile intended to break free from the pull of the Earth, and on technical progress involving motors, fuels and structures. While on this subject homage should be paid to the German Hermann Ganswindt who in 1881, two years before Tsiolkovski published his paper, had recognized the fundamental importance of the escape velocity (11.3 km/s) and had proposed to equip a rocket which would escape terrestrial attraction. Ganswindt therefore enjoys a theoretical precedence to Tsiolkovski, but the significance and the duration of the Russian's work is acknowledged.

The same goal of conquering space spurred the Frenchman Robert Esnault-Pelterie, who, moreover, coined the term 'astronautical' in 1912 and also the American Robert Goddard in 1916. However each of them neglected Tsiolkovski's masterful conclusions. He had tackled the problem at a time when the **thrust of the exhaust gases** was fed only by **solid fuels**, and had deduced justifiably that this would power only an initial thrust and that the load/weight relation of the rocket meant that either a large rocket could go a short distance or else a small rocket a long distance. **Liquid fuels** on the other hand spread out the thrust in a rather more uniform way (see *Great Inventions Through History (Rockets)*). The 1914–18 War did not allow much scope for astronautical projects and the solid fuel rockets sufficed in principle for the requirements of bombing. Their principal use was the **rocket-bomb** invented by Yves Le Prieur in 1916 (see p. 215); it was launced from an aeroplane towards German tanks. Esnault-Pelterie tried, but in vain, to make rockets which would function as auxiliary propellers for aeroplanes.

With the return of peacetime it was again possible to spend time on astronautics. In March 1926 in Worcester, Massachusetts, Goddard launched the first non-military rocket of the 20th century which opened up the Space Age, despite its modest performance. It reached an altitude of 60 metres propelled by a **motor with a compressor and turbine** which ran on **liquid oxygen** and **petrol**, and it was very small (only about a metre long). Three years later, Esnault-Pelterie and the banker André Hirch founded a prize for astronautics. Astronautics firms proliferated. Also in 1929 the German Hermann Oberth (the teacher of Wernher von Braun, designer of the V2 rocket bomb and leader of the American Space Program) clarified further the still vague concepts of astronautics in his major work *Wege für Raumschiffahrt ('The Way of Space Travel')* and was the first to put forward the idea of **multiple fuel reserves** which would eventually lead to **multi-stage rockets**.

Goddard was the first in 1916 to suggest the use of moon orbiting spacecraft to carry out photographic observation. He was also the first to propose the idea of interplanetary propulsion by **ion ejection**, which Oberth was to take up in 1929, adding the possibility of **electrostatic propulsion**. In 1953 the German Eugen Sänger in turned proposed the idea of **photon propulsion**, that is, by the emission of photons in space. In the 1980s these three modes of astronautic propulsion were still at the project stage.

Automatic car guidance system
Philips, Sagem, Renault and French Telebroadcasting, 1985

In 1985 the above four firms made more or less similar systems which were meant to keep the driver permanently informed on the state of his vehicle, from tyre wear and tyre pressure to the information on the best route to follow to arrive at a given destination, taking into account the distance and the road conditions. This latter information is provided by radio. In 1987 the three systems, Carin, Minerve and Atlas were combined to make one, called **Carminat**. Further progress was made in 1991.

Automatic gearbox
Föttinger, 1910

It was in 1910 that the German Hermann Föttinger invented the **automatic gear-changer**, which was made using a torque converter or special drive belt system; when mounted between the drive shaft and propeller shaft, it enabled the suppression of the **clutch**. After that, the gearbox functioned only in three ways, in low gear or town speed, cruising speed and in reverse.

It was this mechanism which several years later would form the basis of development for the automatic gearbox.

Automodulated horn
Sparks and Weatherington Co., 1929

The first car hooter with two or three tones, which was meant to 'personalize' a vehicle, was invented in 1929 by the American car accessory firm Sparks and Weatherington.

Bathyscaphe
Piccard, 1905

The first enclosed, airtight machine which could descend to great depths underwater with men inside it was conceived in 1905 by August Piccard, an American citizen of Swiss origin. It had a float filled with lightweight fuel and was equipped with electric accumulators activating vertical and horizontal propellers. It has given rise to a generation of increasingly improved machines, the most recent prototype of which is the French *Saga* which was launched in 1987.

As predecessors it had the **diving saucers** — to a depth of 3000 m — *Alvin* (United States), *Cyana* (France) and *Pisces* (Canada), and the **exploration submarines** *Nautile* (France), *Trieste* (United States) as well as the recent versions of Piccard's bathyscaphe — to 6000 m.

Cable-car
Anon, United States, or Ritter, 1866

It is difficult to assign a name and a date to the invention of the cable-car; it is even possible that this was only an aerial adaptation of the haulage system of the trucks in the mines which was transposed to work at a height. The cable-car seems to have appeared almost simultaneously in the United States and in Europe. In the United

States it was an anonymous invention and consisted of the towing through the air of small trucks which were used for transporting rubble and coal in the mines; the trucks were suspended from cables and towed up to a cabin. In Europe in 1866 the German W. Ritter set up a developed cableway with four cables and a carrying distance of 101 m, which used a winch for the towing. The cable-car could carry two passengers and the vehicle was designed for the supervision of the hydraulic turbines on the River Rhine.

Car air conditioning
Wilkinson, 1902

In numerous American states in summer, saloon cars quickly become stifling, and so the American car manufacturer Franklin asked his engineer, J. Wilkinson, to make an air conditioning system which could be installed actually inside the vehicle. This took place in 1902; it consisted of a **water-cooling system** similar to that of a conventional radiator, with special air pipes inside the car.

Car anti-theft device
Niemann, 1934

The first anti-theft device for a car was invented in 1934 by the German Abram Niemann; it blocked the steering column.

Car disc brakes
Lanchester, 1902

At first the motor car was equipped with mechanical or hydraulic **brake shoes**. These brakes were problematic in that they lost a fraction of their efficacy after a series of brakings, and it was to counteract this that in 1902 the Briton Frederick W. Lanchester designed disc brakes; these had an excellent **heat dissipation** which considerably reduced the tendency to distort which was the main problem with brake shoes.

They were expensive, and it was only during the 1970s that safety requirements demanded their use in mass production lines.

Car speedometer
Thorpe & Salter, 1902

The London firm Thorpe & Salter put the first speedometer in to commercial production in 1902; it was attached to the rear axle of the vehicle. It was graded from 0 to 35 miles per hour.

Caterpillar tracks
Holt, 1904

The invention of **armoured vehicles** dates back at least to the 14th century (see *Great Inventions Through History*). They did not become widely accepted, for the very reason of the weight of these vehicles, which made them notoriously badly suited to travelling over rough or steep ground. In the 18th century however, it became obvious that the problem with the armoured vehicles as with all heavy vehicles lay in the **weight distribution**: if these vehicles were on wheels, all their weight was carried by the point of contact of the wheels with the ground. In 1770 the Irishman R. D. Edgeworth had the idea of spreading the load by using **wheels with segmented fellies** (rims), the segments of which would be linked together. The outline of the caterpillar track was improved gradually by the Englishman Boydell in 1854, then the Italian C. Bonnagente in 1893. The problem was only rationally and practically solved with the caterpillar track pulled by **driving wheels** which was invented in 1904 by the American Benjamin Holt and completed a short time later by his compatriot C. L. Best. The first set of caterpillar tracks was designed for use on an **internal combustion engine tractor**.

Deltaplane
Wanner, Rogallo, 1948

The Deltaplane, which is derived from the **glider** (see p. 197), was invented in 1948 by the American Francis Melvin Rogallo (whence came the machine's name at first, **Rogallo wing**). It was a triangular wing of woven metal wire coated with silicone film which was used for **hang-gliding**. The invention may have been made some 10 months earlier by another American, Wanner.

Ejector seat
Junkers, 1938

The idea of an ejector seat circulated in international aeronautical circles from the beginning of the 1930s. The origin of it is difficult to establish however, and the distance between the idea and the realization, difficult to estimate. It seems that the invention should be attributed to the German aeronautical firm Junkers, who tested ejector seats in ground vehicles in 1938, the ejection being effected by a **compressed air gun**. The first aeroplane to be equipped with one was the Junkers Ju-88 in 1939, but the ejector seat was not properly adopted by the Luftwaffe until 1941.

Electric railway
Cazal, 1864; Siemens, 1878

The electrification of the railway lines was unusual in that it followed soon after the arrival of the railways themselves but took nearly a century to dominate, because of the longstanding rivalry between coal and electrical energy. From the beginning it had definite advantages as a means of urban transport.

At the end of the 18th and beginning of the 19th centuries, the horse-drawn tramways, individual carriages and commercial coaches created intolerable traffic jams in all the large cities of the world. The arrival of the locomotive aggravated the problem by adding unbearable air pollution, which led the town councillors of New York, London and Paris to envisage an **underground railway** (see *Great Inventions Through History*). The solution rested obviously with electric energy which would end the locomotive smoke and enable the installation of viable underground railway lines (the first underground engines were designed to emit their fumes into the water of their boilers in order not to make the air in tunnels unbreathable). In 1864 for the first time the Frenchman Cazal transferred the energy from a magneto-electric motor to the axle of a locomotive, simply following the example of the Russian B. S. Jacobi who had transferred the same energy to a propeller axle in 1834. But the first to build an electric motor capable of pulling an engine was the German Werner von Siemens in 1878. It consisted of a small motor unit which was powered by a **continuous current** passing through a **third rail** placed between the two traditional rails and was demonstrated at the Berlin Fair in 1879.

The first electric railway was opened at the Giants' Causeway in Ireland in 1884; before electric locomotives could compete with steam ones on a large scale they had to produce equivalent power.

Escalator
Reno, 1892; Wheeler, 1892; Seeberger, 1899

Two kinds of moving stairway were invented. The **Reno Inclined Elevator** was patented by Jesse W. Reno in 1892, and installed at the Old Iron Pier on Coney Island in 1896. It consisted of a continuous belt conveyor made up of wooden slats; it was driven by an electric motor and travelled at a speed of about 1½ mph. A moving staircase with flat steps was patented by Charles A. Wheeler in 1892; the patent was bought in 1898 by Charles D. Seeberger who made an improved design. This was built by the Otis Elevator Company; it was installed at the Paris Exhibition in 1900 and re-erected the following year at Gimbel's Department Store, Philadelphia. Seeberger gave the name **escalator** to his device. It incorporated the **combplate landing device**, which was a feature of the Reno Inclined Elevator, and which enabled passengers to step off the end of the escalator.

The first moving stairway in the UK, a Reno Inclined Elevator, was installed at Harrods, London in 1898; the first escalators connected the District and Picadilly platforms at Earls Court Underground Station, and were opened to the public in 1911. From the 1920s escalators became common in the underground stations and in department stores.

A new variation of the invention was the spiral escalator, the first of which was installed in Japan in 1985 by Mitsubishi Electric Corporation.

Front-wheel drive
Grégoire, Fenaille, Nugue, 1926

The idea of a car with the driving wheels in front (meant to ensure better road holding), was the brainchild of Jean A. Grégoire and Pierre Fenaille in 1926 and was made possible due to the work of the engineer Nugue. Contrary to certain assertions, the first front-wheel drive vehicle which had a **universal joint**, also called a **homokinetic joint** and a **Cardan joint**, was Grégoire's Tracta. It was on the road at the beginning of the summer of 1926 and was successful in the 'Le Mans Twenty-Four Hour Race' in 1927. The principle of front-wheel drive was successfully adopted in 1929 by the German firm D.K.W. and Adler, and by the American Ford, then by the Belgian manufacturers Astra and Juwel, as well as the French manufacturer Donnet in 1931. The prototype made by the Bucciali brothers must also be mentioned, which was built in the same year, 1931. It was not until 1934 that the front-wheel drive car reached mass production with André Citroën's famous 7A.

Gliding
Le Bris, 1856; Mouillard, 1865; Lilienthal, 1896

Gliding has developed from the observation of the flight of birds, which dates back at least to Leonardo da Vinci in the 16th century. It is possible that the Chinese attempted to fly by floating in the air on a **kite** (see *Great Inventions Through History*) but this was not a proper free-gliding flight.

The first to try gliding seems to have been the Frenchman Jean-Marie Le Bris in 1856, who managed to take off from a Breton beach on a **glider** which he had made. In 1865 his compatriot Louis Mouillard experimented with flying in Algeria and succeeded in making a flight of about 40 metres a short distance from the ground. Author of works such as *The Empire of the Air, Experiments in Applying Ornithology to Aviation*, and *Flight without Flapping*, Mouillard seems to be the true originator of long-distance gliding. Much research ensued concerning experiments with flight; the most famous is undeniably that of the German Otto Lilienthal who attempted a take-off on jointed wings in 1896 and killed himself when he crashed to the ground. It was not however until 1920 that gliding began to spread, particularly in Germany. In 1939 there were 200 000 pilots of patented gliders on record. This enthusiasm had a military motive: it was with gliders that the Germans took the fort of Eben-Emael on the Albert Canal, and then carried out their landings in Crete. They used gliders for military purposes until 1945. The British used gliders to transport personnel for the Normandy landings in 1944.

Gliding as a sport was launched in 1922 by the Frenchman Pierre Massenet.

A new development in the 1970s was the autonomous take-off hang-glider, following work by the inventor Dave Kilbourne.

Helicopter

Bréguet brothers, 1907; Cornu, 1907; La Cierva, 1923; Focke, 1936; Sikorsky, 1944

A helicopter is a craft heavier than air which is both kept aloft and propelled forward by a **vertical axle propeller**. The actual principle of it is very old (see *Great Inventions Through History*) but apart from the one in which the Italian Enrico Forlanini succeeded in taking off to a height of 15 m for nearly a minute, using a tiny steam engine which activated two co-axial propellers which were mounted on the same axle and rotating in opposite directions, the 19th century ended without further development. It was in 1907 that, inspired by the Frenchman Charles Renard's work on propellers and using an internal combustion engine, the brothers Louis and Jacques Bréguet achieved the first piloted flight in a helicopter which had four rotors comprising

32 aerofoils. Paul Cornu's helicopter, which was less complicated since it had only two rotors, was built in the same year but it did not resolve the problems of direction encountered by the Bréguets' machine. The Wright brothers' first successes with the aeroplane tended to discourage research on helicopters. Between 1908 and 1912 however, the Russian Igor Sikorsky managed to improve its **directional stability**. Then, between 1919 and 1923 the Italian Pateras Pescara, the Germans Emil and Henry Berliner in the United States, and the Frenchman Georges de Bothezat, also in the United States, designed machines which, without using any fundamental new invention, were good enough improvements to make flights of nearly an

The helicopter designed by Igor Sikorsky. *It reached 70 km/h, took off vertically, but needed 30 metres of space to land. This was the last of the Sikorsky single-seaters: the model which followed could carry five people at a speed of 160 km/h.*

hour at low or medium altitude, with between one and three passengers. These machines vibrated a great deal however, and they were not very reliable.

The really acceptable improvement was brought about in an incidental way by the Spaniard Juan de La Cierva. In moving away from the fundamental principle of the helicopter, which is that the same propeller both holds the machine aloft and propels it forward, that same year La Cierva made an **autogiro**, an aeroplane without wings which takes off with the help of an ordinary aeroplane propeller. As the machine increases its speed, a horizontal propeller with **freely revolving blades** starts moving which acts as an aerofoil which enables the machine to take off. The autogiro did not have many successors, even though the principle of free-moving blades was used in certain experimental machines during the 1950s, such as the Fairey Rotodyne, but it introduced a first-rate technical novelty: the blades of the rotor were not fixed onto the shaft but jointed, which avoided the vibrations caused by the rigid blades. Moreover, since the blades themselves were flexible they adapted much more easily to the weight distribution and cycle changes caused by the increase in speed. It therefore became possible to travel more quickly.

New progress in 1936: the German Heinrich Karl Focke invented a device which altered the **angle of incidence of the blades**, allowing it to fly equally well backwards as forwards. Paradoxically, however, no one then realized the importance of this essential improvement to the modern helicopters. It was only in the 1940s that the specific characteristics of the helicopter began to take shape. This craft is not a competitor to the aeroplane; its ability to hover, even at a few centimetres above the ground, to go backwards and land in extremely small spaces makes it into a kind of 'all-terrain vehicle' which is ideal for fire fighting in inaccessible places, mountain rescue, road surveillance, some reconnaissance missions, low flying, and the transportation of injured people or other loads in steep or difficult places as well as being an essential part of modern warfare.

It was only in 1944 that the above-mentioned Sikorsky in the United States, taking into account all the progress already made, built the first modern helicopter. The VS 36A had an enclosed cabin containing the cockpit and was equipped with a lattice-work tail at the end of which a perpendicular propeller on the axle of the rotor served to prevent the fuselage from turning over on itself; the blades of the rotor were not only of **variable speed**, which meant that the speed could be increased independently of the axle, but also of variable incidence, which meant that they could be angled forwards or backwards, as in Focke's invention.

Two main derivatives of the helicopter emerged during the 1980s: the **vertical take-off aeroplanes** or **VTOL**, based on the principle of a modification in the angle of incidence of the propellers which is very pronounced, and **flying platforms**, of the American Hillier type, which are helicopters with streamlined propellers for transporting an individual in the open air at a low altitude and over short distances.

Hovercraft

Dornier, 1929; Cockerell, 1953–9

The idea of a maritime vehicle which would travel on a **cushion of air** was first thought of by the English engineer John Thornycroft, who formulated the theory that the resistance to forward movement could be reduced considerably if the hull had a concave shape and captured a mass of air underneath it which would act as **shock-absorber** and **slide** at once. Thornycroft patented this invention in 1877 but never managed to build a boat to demonstrate its reliability. Several engineers subsequently tried to do so, and the first conclusive attempt to remain on top of a cushion of air, although the craft used was not specific, was that achieved by the hydroplane *DO X* made by the German firm Dornier in 1929. Indeed, this machine increased its performance notably during a crossing of the Atlantic by flying skimming the waves (a technique subsequently used by reconnaissance planes during the World War II).

The Briton Christopher Cockerell featured among the several engineers who took up Thornycroft's principle, but with the intention this time of creating a specific craft. His studies, beginning in 1953, were aimed at the creation of an amphibious vehicle; this would have an inverted hull which would be kept aloft by an airflow from above, created by a horizontal propeller. Cockerell found that the performance was considerably improved using the same size of motor if the air was pumped, not into the whole hull, but only into a curtain which was fixed around the circular hull. Therefore the principle called the **open chamber** was succeeded by that called the **peripheral jet**. This invention was patented on 12 December 1955 and the following year Cockerell founded Hovercraft Ltd.

The first ever hovercraft was launched in the utmost secrecy under the name *SRN1* in November 1959, with the collaboration of the British Minister of Defence. It weighed 4 tonnes and could carry four men at 25 knots an hour. Cockerell came up against the same problem as Thornycroft, the difficulty of maintaining a cushion of air under pressure, since the air tended to escape towards the outside despite the centripetal force against it. He had equipped his craft with supple **skirts** made of thick rubber material, which kept the air imprisoned under the hull. The skirts were improved and were made out of a special type of plastic, which allowed the speed of this hovercraft to be doubled and the weight of the engine to be increased to 7 tonnes. The propeller was powered by a **gas turbine**.

The hovercraft raised hopes which to a certain extent exceeded its capacity. It had in fact been thought that the hovercraft might become a form of land vehicle, which would be capable even of travelling over roads. But right up until the end of the 1980s the imprecision of their directional system excluded the possibility that they could find a use analogous to that of motorized land vehicles. Moreover, the use of gas turbines in amphibious or maritime vehicles includes a serious inconvenience: the vaporization of sea water in the ducts which causes rapid corrosion. Using **diesel motors** in place of turbines has remedied this but at the price of making the engine considerably heavier. Despite these drawbacks and some others, active research continues which aims to improve the efficiency, profitability and upkeep of hovercraft.

Cockerell made his first experiments on air cushion flotation using tins of canned food. He was such a brilliant theoretician, and had gone into the theory of air cushion propulsion in such depth that more than quarter of a century after the first hovercraft had been launched, all his ideas had still not been analysed.

Hydrofoil

Forlanini, 1900

A hydrofoil is a boat with wings placed under the hull, both behind and in front; these wings are fixed in place by vertical spars. When it is stationery the hydrofoil resembles an ordinary boat, its wings being under water; as the boat accelerates the **lift effect** created by the wings raises the hull out of the water. The speed increases due to the fact that the only resistance to forward movement is that of the side frames, the front-facing edges of which are stream-lined. The first craft heralding the hydrofoil was that built in 1897 by the Count of Lambert; it consisted of a **catamaran**, the two parts of which were linked by four rigid wings. It is not known though if the 'hydroplanes' effectively used the lift-by-acceleration principle of an actual hydrofoil. The first apparatus of this kind without confusing features was the one built by the Italian Enrico Forlanini in 1900 (the principle of it had been established by him in 1898); in 1905 it reached a speed of 80 km/h

which was remarkable for the time. Since then engineers and mechanics showed a great deal of interest in hydrofoils and in 1918 the Briton Alexander Graham Bell, the inventor of the telephone, and his compatriot Casey Baldwin broke the record at 100 km/h on a hydrofoil equipped with two 350 cc motors. Several hydrofoils are used throughout the world for regular, short-distance sea links and they come in different types.

The speed of the hydrofoil which is about 80 km/h (50 knots) has its limits fixed by the **cavitation** effect, which is the formation of bubbles of air caused by the turbulence of the water. Indeed, as the bubbles burst, or explode to be more precise, they very quickly break down the surfaces of the air lift and the high speed causes the air cavities which form under the wings to fill up with water, thus causing the air lift to be reduced.

The Lambert hydrofoil, built in 1897 and equipped with a Renault 280 cc engine.

Hydrogen aeroplane
Hans von Ohain, 1935; NASA, 1956–9

An aeroplane whose propulsion would be effected not by the combustion of a hydrocarbon (kerosene), but by combustion of hydrogen, was the subject of study by the German Hans von Ohain, as well as several engineers from the German firms Messerschmidt and Heinkel. This study was to lead to the Messerschmidt Me-163 VI *Komet*, which was used in the World War II (see **Aeroplane**). The idea was taken up again in 1956 by NASA (the National Aeronautics and Space Administration), which modified a twin jet engined B57 bomber by equipping it with tanks of liquid hydrogen at the ends of the wings. In 1958 a Curtiss-Wright J 65 managed a series of experimental flights during which it used liquid hydrogen and conventional fuel alternately.

Research in this direction was abandoned in 1959, firstly because of the **high cost** of liquid hydrogen, which is nearly three times that of kerosene, and the **risk of explosion** which is somewhat higher than with conventional fuels. Liquid hydrogen has to be stored at –252.5°C. Nevertheless, in 1897 the Soviet Andrei Tupolev, the famous founder of the aeronautics firm which was named after him, modified a conventional apparatus, the Tu-155, which succeeded in its experimental flight on 15 April 1988. At the same time, the Americans were preparing for the launch of a hydrogen-powered craft, the National Aerospace, which was nicknamed the Orient-Express because of its potential to link Washington to Tokyo in two and a half hours. One of the new generation of hypersonic planes, its launch is planned in the mid-1990s.

> One of the theoretical advantages of the hydrogen aeroplane is that it does not produce oxides of carbon, and therefore makes less pollution in the atmosphere.

In-flight refuelling

The first in-flight refuelling of one aeroplane by another was carried out in 1929 by the US Army Air Corps, now the US Air Force, when a Douglas C1 transferred some fuel to a three-engined Fokker using a pipe.

Subsequently two refuelling planes were to transfer 5000 gallons of fuel to the Fokker, in several stages, as it was taking part in an endurance test which meant keeping it in the air for seven days.

Internal combustion engine
Beau de Rochas, 1862

Induction, compression, power, exhaust. These four stages of the **four-stroke cycle** heralded the modern age of the automobile; they describe the internal combustion engine. In order to understand the invention of it, one must return to Huygens' principle. At first sight it could seem that it was the internal combustion engine that Huygens designed; but in fact it was atmospheric pressure which drove the piston downwards and this is why this machine was called an **'atmospheric engine'**.

Nothing approached the actual internal combustion engine until the invention by Philippe Lebon, who was the first to think of introducing an air–gas mixture into a cylinder, which is compressed, and ignited by an electric spark, thus generating the expansion which pushes the piston. Philippe Lebon was able to study neither its manufacture nor its improvement for he was mysteriously murdered three years after his invention; in the meantime he had invented **electric ignition** and **compression**. After him several inventors tried to build an engine using gas and internal ignition: the Englishman William Cecil (1820) and W. I. Wright (1830); the Italian Luigi Cristoforis (1823); then there was another Englishman, Samuel Brown, who is the only one to have completed his research, and the Italians Eugenio Barsanti and Felice Mattecci (1854).

The real theoretician of the four-stroke cycle was Alphonse Beau de Rochas; without this the internal combustion engine would have had a much weaker output. Beau de Rochas proposed four strokes for an internal combustion engine instead of the two of a steam engine. However Beau de Rochas did not carry his invention through to a successful conclusion, due to problems with the proposed ignition of the air–gas mixture.

In 1873 the Englishman George Brayton, to allow air–gas mixing, advocated placing the suction–compression demi-cycle outside the cylinder, which would have the advantage of maintaining a constant pressure throughout the whole admission and ignition stage, with the suction and compression taking place in an isolated tank, an accessory anticipating both the carburettor and the compressor. Brayton did not succeed, because of persistent problems with ignition, but on the other hand there were two great achievements: his idea was going to serve as a base for research into **turbomotors**, and, above all, Brayton had been the first to propose replacing the air–gas mixture with liquid fuel.

The technological progress was going to be able to continue and it did progress quickly. In 1878 the German Nikolaus August Otto presented an internal combustion monocylinder engine at the Universal Exhibition in Paris. It worked satisfactorily and had a cycle almost the same as the one invented by Beau de Rochas: 1 — suction of the mixture during one whole stroke of the piston; 2 — compression at the return of the piston; 3 — ignition at the maximal point of compression, or dead point, then combustion; 4 — expulsion of the burnt gases during another stroke of the piston. The connecting rod of the piston was attached via a crank to a flywheel which controlled the inlet valves for admission of the mixture and the exhaust valves for the expulsion of the fumes. The Otto engine was immediately successful: more than 35 000 of them were made in a few years. These were large fixed engines (600 horsepower), and very powerful. And they were still only gas engines, requiring a nearby boiler for the supply of gas from coal or oil.

The petrol internal combustion engine was still to come. It appeared gradually, firstly because liquid fuels were distrusted, and then because the factories which produced gas for the new motors began to accumulate large quantities of by-products such as fuel oil and petrol. In 1879, the Englishman Dugald Clark showed that it was possible to reduce the four-stroke cycle of Beau de Rochas' engine to a two stroke: one for the suction, compression and ignition, another for the expansion and expulsion of the fumes. In 1880 the German Carl Benz used this as a basis for experimentation on a light engine, but without success. Then the decisive step was made with an invention by another German, Werner von Siemens, who in 1884 developed **low-voltage electrical ignition**. After that there were fewer risks of explosion and both four- and eventually two-stroke petrol engines. It was a four-stroke petrol engine of a two-cylinder type designed by Murnigatti which, for the first time, the German Gottlieb Daimler positioned on a machine with wheels, a wooden motorbike. This had also been equipped with a new addition, the **carburettor**, for mixing the petrol with air.

The Otto patent was revoked in 1886, when Beau de Rochas was able to establish the precedence of his invention and the near-identity of the Beau de Rochas cycle and the Otto cycle.

Laminated glass windscreen

Corning, Saint-Gobain, 1929

During the 1920s automobile accidents drew manufacturers' attention to the danger to the face if the **windscreen** shattered. So the large glass manufacturers researched a type of glass which would not fly into pieces under impact. The American firm Corning Glass discovered that glass which was rapidly cooled after being heated had a much tougher layer on each side than that of ordinary glass. Moreover, this glass lent itself more to **streamlining**, given that the manner of heating gave it a much greater ductility than that of ordinary glass. Under impact it broke into small rounded pieces and not into large sharp splinters. This is what was called **toughened glass**; in the United States it was not authorized for sale because of its hardness, which is greater than that of the skull. It was necessary to turn to another type of glass, which at the beginning consisted of a sheet of **celluloid** between two layers of glass, called **laminated** or **layered glass**, which was much less hard. It was made in France in 1929 by Saint-Gobain and presented much less risk for the motorist. A few years after it had started to be produced it nevertheless fell out of favour, for in certain conditions the celluloid changed and became opaque. Laminated glass only became popular again when the celluloid was replaced by a derivative of **vinyl**, which was much less susceptible to distortion.

Magnetic levitation

Bachelet, 1912

Magnetic levitation is the name for the supporting of an object in air without any point of contact by the effect of magnetic repulsion between like poles of a magnet. The principle is sufficiently viable that high-speed magnetic-levitation trains were being developed in Germany, Japan and the United States in the 1980s. For example, a train with a base formed from the North pole of a magnet a few millimetres away from a rail also formed from a North pole; by eliminating the friction, speeds of around 400 km/h can be attained.

The principle in question was stated in 1912 by the Frenchman Emile Bachelet, in the light of research on **superconductivity**. Indeed with this enabling the induction of electric currents of considerable strengths in certain metallic composites, it became possible for equally powerful magnetic fields to be generated.

Microwave aeroplane

Canadian Communication Research Centre, 1987

A prototype of an aeroplane without an automatic motor was tested successfully in Canada in 1987. It had a 4 m wingspan, $1/8$th of the width of the planned aeroplane, and was equipped with an electric motor powered by a high-intensity beam of micro-waves, which were transmitted by a parabolic antenna on the ground. The microwaves are captured by receivers placed under the aerofoils and under a parabolic antenna; these receivers are **diodes**. The craft is meant to assure a localized road and ecological photographic surveillance, within a radius of about 300 km, and to serve as a **radio broadcasting relay station** by replacing ground networks. Designed in 1982 and made by the Canadian Com-munication Research Centre, this pilotless aeroplane can fly indefinitely and complete the satellite network at very small cost. The real cost of the completion of this apparatus was estimated in 1987 to be $13 million.

Motor car

Daimler and Maybach, 1889; Panhard and Levassor, 1891

There is uncertainty over the absolute inventor of the first motor car, that is, one which used **petrol** as fuel. The immediate precursor seems to be the **quadricycle** which was built in 1889 by the Germans Carl Benz and Wilhelm Maybach. This was based on a **tricycle** propelled by a **two-stroke monocylinder motor** which was made by Carl Benz in 1883. The motor of the quadricycle turned over at about four times that of the tricycle. If these really are the first two vehicles powered by an inter-nal combustion engine and petrol, the first car which consisted of a chassis — tubular — with a motor in front was that which was built in 1891 by the Frenchmen René Panhard and Emile Levassor, following the Daimler–Maybach quadricycle for which they held the licence. The firm Panhard–Lavassor was also the first to mass-produce cars.

To satisfy certain historians, mention must be made of the car built in 1886 by the Danish firm Hammel and which is still in working order.

It could be said that even though the

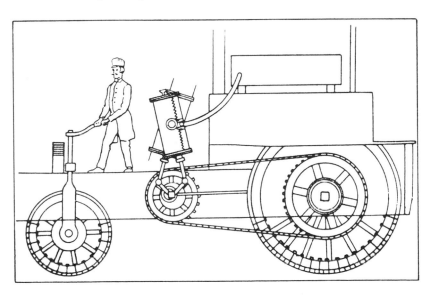

The extraordinary machine is a steam-powered vehicle with two rocking cylinders which was designed by Charles Dietz in 1835. Dietz belonged to a family of engineers and succeeded in building vehicles like this which are mid-way between steam engine and automobile.

motor car subsequently benefited from numerous additional inventions, as well as endless improvements such as the **tubular** **chassis, steel framework** or **monocoque construction**, the original invention remained fundamentally unchanged.

When the American George B. Selden patented a petrol driven motor car in 1879, a long dispute took place with Daimler–Maybach over the precedence of the invention. Selden never built any cars and his name is retained in history only in an incidental way, as the originator of a 'trick' which verged on fraud: he claimed to have the rights of any motor vehicle built anywhere in the world, and actually succeeded in this until 1903. The outcry against Selden's claims, which until then were upheld by American law, began in France where Panhard and Levassor, then Jeanteaud, who were quickly joined by Henry Ford, examined all the patents previous to that of Selden and instituted proceedings which ruined him completely. Indeed, his patent had only been granted on 5 November, 1895, under the number 540 160. It is undoubtedly one of the most startling examples of abuse authorized by the patenting system.

Muscle-powered aeroplane or aerocycle

Nieuport, 1921; McReady, 1979

As a theoretical concept the muscle-powered aeroplane is certainly very old and dates back at least to Leonardo da Vinci in the 15th century, if not to the myth of Icarus. But it is just as true that the technical data which could enable a craft heavier than air to take off and move forward solely by muscular power had not begun to be studied until the 20th century. The pioneer of this was the Frenchman Edouard Nieuport, who in 1921 put his knowledge as competitive cyclist and aeroplane builder to good use to make a flying bicycle or aerocycle. Christened *Aviette*, the apparatus was 'piloted' by Gabriel Poulain; it rose to a height of 1.50 m and travelled a distance of 10.54 m at Longchamp racetrack in Paris, thus taking the Peugeot prize. Several versions of the aerocycle ensued during the following decades but they did not manage to get it widely accepted as a useful means of transport. It was the American engineer Paul McReady who tackled the problem of construction and incorporated new ultra-light materials, and was the first to prove that the aerocycle could display a certain degree of reliability.

In 1979 an aerocycle made of **carbon fibre** and **synthetic fabric**, the *Gossamer Albatross*, piloted by Brian Allen, crossed the 22 statutory miles of the Channel in 2h 49min. The propeller, as in Nieuport's *Aviette*, was worked with pedals by means of a gearing system. It seemed at the end of the 1980s that the aerocycle's lack of stability, the hazards of the effort required to keep it going and the time it took, and finally the cost of the apparatus placed it in an inferior position in relation to amateur flying crafts such as the **ULM** and the **Deltaplane**.

Pressurization of aeroplanes

Boeing, 1938

The maintenance of **atmospheric pressure** in aeroplanes, which is meant to compensate for the low pressure at high altitudes, is more of an innovation than a proper invention, but it has been so important in the evolution of aviation that it deserves to be mentioned. The first aeroplane to be pressurized, using a compressor, was the Boeing 307 Stratoliner, which was built in 1938 by the American firm Boeing and brought into service in 1940. Previously the passengers, essentially the crew, of aeroplanes which flew above 3000 m were equipped with **oxygen masks**.

Pressurization, which maintains the pressure in the cabin to the equivalent of atmospheric pressure at 2500 m, has enabled commercial aeroplanes to fly at altitudes of 10 000 to 12 000 m and therefore be less affected by turbulence at lower altitudes.

Rear-view mirror

Cockerill, 1896

With the advent of the motor car came the need for the driver to be able to check behind him, without turning round, to see if he was being followed by another vehicle travelling at a faster speed. It was this that led the English military surgeon John William Cockerill to have the notion of placing a mirror above the windscreen, which when correctly angled would provide the necessary information.

Retractable undercarriage

Wiencziers, 1911

The retractable undercarriage was invented in 1911 by the German aeronautical firm Wiencziers. It was used to equip this manufacturer's monoplanes from that year onwards but took some 30 years to become widespread. It was the growing **speeds** and **aerodynamic** considerations which caused it to become universally used.

Rickshaw

Scobie, 1869

One of the most unusual inventions in the domain of transport is the two-wheeled car which is pulled by a human. This is called a rickshaw and is very widely used in Asia. It is believed to have been invented in 1869 by the American missionary priest Jonathan Scobie, to transport his invalid wife in the streets of Yokohama in Japan, in such a way that she could use the alleyways and park without difficulty. The more modern version is powered by a bicycle.

Scooter
Auto-Ped, 1915

The ancestor of all scooters, or two- or three-wheeled **streamlined** and **motorized scooters** was made in 1915 by the American firm Auto-Ped Co. It was not a success and it was not until after the World War II that Italian industry managed to get this mode of transport accepted, as a two-wheeled vehicle.

Solar-powered aeroplane
McReady, 1980

The *Gossamer Penguin* was built by the American scientist and aviator Paul McReady. It was equipped with a motor powered by **photovoltaic cells** (which generate a voltage when light falls on them), themselves powered by **solar energy**, and achieved a flight of 3.028 km at an altitude of 4 m in 14 min in the Mojave desert in 1980. The pilot was Janice Brown. The following year another version, the *Solar Challenger* which had 16 000 photovoltaic cells, made a 368 km flight crossing the English Channel at an altitude of 11 000 feet in 5½ hours.

Traffic lights
Knight, 1868; Benesch, 1914

The first known traffic lights were for pedestrians who had to cross main roads or walk along the tramways; they actually consisted of a **semaphore signal** which raised a red glass and a green glass **gas lamp** alternately; only one example is known to have been installed, which controlled the traffic in front of the House of Commons, but it only stayed there for a few months because one of the lamps exploded, killing the attendant officer. J. P. Knight, the signal engineer of the English railway system who had invented it, did not return to the project. Traffic lights did not come into widespread use until after 1914 when the American Alfred Benesch had completed their automation. The first town to be equipped (with a red light) was Cleveland, Ohio, in 1914. It was not until 1918 that the modern **tricolour system** incorporating an amber warning light was set up in New York.

Turbojet
Whittle, 1928–30; Vickers, 1940; Rolls-Royce, 1945; Wibaut, 1953

A turbojet is the name for a **jet engine** in which the air let in at the front is pushed into a **combustion chamber** where fuel is injected; the combustion generates gases whose thrust is greater than that of the incoming air and the ejection of which creates the propulsion, usually of an aeroplane. The turbojet has been used mainly in aviation. The theory behind it was published in 1928 by the Briton Sir Frank Whittle, who patented it in 1930. The German Hans von Ohain in turn patented a similar principle of propulsion in 1935, and it was in Germany in 1939 that the first flight took place using an aeroplane with a turbojet engine, a Heinkel He-178. Research in Great Britain and the United States was then vigorously speeded up. In 1941 the British tested the Gloster Whittle E-28, which was also equipped with a turbojet; it led to the manufacture in 1944 of a craft with a much better performance, the Gloster Meteor. The Luftwaffe replied by bringing out the Messerschmidt Me-262. The speed and reliability of this type of engine was to assure its supremacy in civil and military aviation right up to the present day: it is Olympus turbojets that power Concorde, for example.

In 1940 the English firm Vickers invented the **twin-flow turbojet** which sends air which has simply been compressed into the end nozzle. This is less noisy and more economical than the previous one and has been used in the Boeing 707, 720 and 727 as well as the Caravelle and the DC-8. In 1953 the Frenchman Michel Wibaut invented the **lifting turbojet** with adjustable nozzles, which effects a transition between vertical and horizontal flight, and which has enabled the development of fighter aircraft which can land and take off vertically.

The **turbojet with afterburner** was invented in 1945 by Rolls-Royce. This has a second combustion chamber behind the first one and so has additional thrust. Since it has higher fuel consumption it is only used for very short periods of time, such as during take-off for civil aircraft and to attain very high speeds for fighter planes (see also **Aeroplane**).

Variable speed propeller
Hele-Shaw and Beacham, 1924

When the variable speed propeller, one of the most remarkable inventions in the domain of aviation, was presented in 1924 by the British inventors H. S. Hele-Shaw and T. E. Beacham, the British Air Minister was only mildly interested. He ordered a dozen to be made and then did not even follow the experiments. The true value of this type of propeller, which enables a maximal return in terms of speed for a fixed number of revolutions, and enables the speeds at take-off and landing to be varied, was not recognized until 1937.

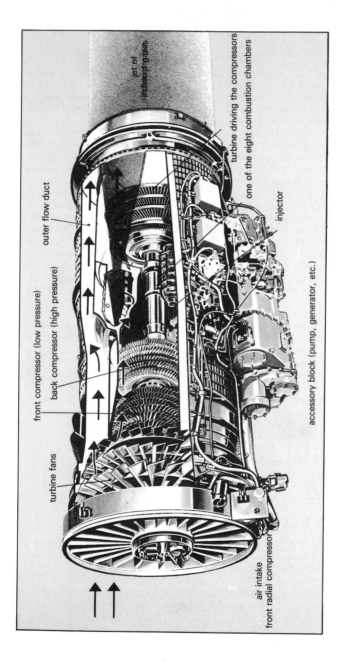

air intake
front radial compressor
turbine fans
front compressor (low pressure)
back compressor (high pressure)
outer flow duct
jet of exhaust gases
turbine driving the compressors
one of the eight combustion chambers
injector
accessory block (pump, generator, etc.)

A Pratt and Whitney bypass turbojet, *with two compressors. The air intake in to the engine is compressed to make an efficient combustion process.*

Wankel engine

Wankel, 1951–4

The Wankel engine is a **rotary engine** with an approximately **triangular rotor**, in which a fixed rotating toothed piston turns during the circular movement of the rotor. Each time one side of the rotor passes in front of the inlet hole, a mixture of fuel enters the chamber of the cylinder. The movement of the rotor compresses this mixture, which ignites once in each cycle. The expansion of the ignited gas exerts a thrust which then makes the rotor turn.

Several engineers had conceived this engine in theory; it was first made by the German Felix Wankel, who worked on it from 1951 to 1954 in collaboration with the firm NSU Motorenwerk AG at Neckarsulm.

After being tested in 1957, the Wankel engine was used in the manufacture of cars by the Japanese firm Mazda from 1967. Though compact, light, and 10% more economical than the classical internal combustion engine, it was never widely used in the whole of the automobile industry (when tested on a small series by Citroën in 1969, it was abandoned due to persistent problems with the **watertightness** of the rotor; also the actual rate of fuel consumption was far greater than in the theoretical calculations). The Wankel engine, which has the ultimate practical advantage of being much more compact than the classical internal combustion engine, in fact represents one of

The Wankel rotary engine, which inspired great enthusiasm among the manufacturers during the 1960s, but which fell out of favour due to intractable problems with its watertightness.

the dead-ends which are plentiful in the history of inventions. Had it been invented at the beginning of the history of the motor car, it would undoubtedly have been successfully developed. But it appeared at a time when the internal combustion engine had already been much improved; it would have required investments out of proportion to its value, so that it was unable to gain the support of the manufacturers. It is possible that the improvement of solar cells will lead one day to the abandonment of the internal combustion engine just as the Wankel engine was rejected for the ecological reasons which are now increasing in importance at the end of the 20th century.

Warfare and
Military Technology

The completion of the atomic bomb and then thermo-nuclear bomb during the decade 1945–55 has initiated a crisis of conscience within worldwide scientific circles which, to this day, the political powers of the large countries have not managed to abate. The proliferation of small countries capable of equipping themselves with nuclear arms has intensified the uneasiness, which has spread the general public. It is impossible to evaluate the effects of this uneasiness, but it seems likely that it has slowed down the invention of means of destruction.

However, it is to be noted that, apart from one or two exceptions, armaments have hardly benefited from any major inventions at all, and the innovations that have ensued since the Second World War are in fact technical, generally civil, improvements on previous inventions. The minicomputers fitted onto missiles or military satellites which are capable of interventions in realtime are one of the dominant features of military technology of the 1980s but they cannot be labelled as inventions in the ordinary sense of the word.

Moreover, military inventions have an increasingly short life span, as witnessed in teh case of the swing-wing aeroplane, which in its time was considered a leading invention but rapidly lost the support of those in command.

In contrast, therefore to the events of the preceding centuries, no inventions from 1850 to the present day have made their mark in any particular development in the art of killing people, excepting the ultimate weapon of course, 'the Bomb'.

Aerial bombing

Curtiss, 1910; Anon., Italy, 1911; Rougeron, 1936

In theory, the origins of aerial bombing can be taken back to 1849; in this year, during the siege of Venice, the Austrians dropped 'pilotless bombs' on that city. These consisted of small balloons of hot air which each carried **a fused bomb** of about 15 kg. These bombs caused only a few casualties and a small amount of damage.

Since the beginning of aviation the potentialities of aerial bombing have grown in likelihood, and in 1908 Count Ferdinand von Zeppelin equipped his aircraft with an armament system comprising five 110 kg bombs. It was not Zeppelin, however, who verified the realiabilty of aerial bombing (the first air raid carried out by Zeppelin was that of an L3 model above the British town of Great Yarmouth in 1915; there were two victims), but it was the American, Glenn Curtiss, who, on 30 June 1910, after practising with dummy bombs, dropped a real bomb on a target in the form of a boat.

Not long afterwards, the Royal British Flying Corps equipped some of their planes with **bomb-launching tubes**. The first real bombing was accomplished by the Italians during the Italo-Turkish War in 1911. An Italian pilot on a Bleriot XI dropped three heavy grenades on targets in Tripolitania.

Aerial bombing became commong practice during the First World War; it was performed by sight and by hand, the bombs often having their pins taken out like hand grenades before being dropped. The damage caused began to be considerable, and the fighters themselves took it for granted that some planes should have been designed as **bombers**.

In 1936 the French theoretician, Camille Rougeron, published the first article on the technique of **dive-bombing**, which had been tested empirically during World War I and then studied by the Americans during the 1920s. The advantage of this technique, which was adopted by, among others, the Japanese military airforce, was that the formations of bombers could approach their target at a very high altitude and so remain undetected; at a short distance from the target they would dive down on to it, creating a big enough surprise to fend off the possibility of an anti-aircraft counter-attack. This technique, however, required a new type of lightweight bomber, with very good handling and a medium range of action.

Air-to-air missile

Wagner, 1943

The first known air-to-air missile was a **self-destructing projectile** which was designed to be launched from one aeroplane towards another; it was invented and built in 1943 by the German, Herbert Wagner. It was designed to be launched by the Dornier Do 217 or Focke-Wulfe 90 aircraft in the direction of HS 298. It was to inflict relatively slow-acting but considerable damage on the enemy's bombing equipment.

A modern development is the **air-launched cruise missle** (**ALCM**), which flies on wings, and is propelled by an engine using air from the atmosphere as an oxielant to mix with the fuel it carries.

Armour plating (of land vehicles)

Hannibal (?) 2nd Century BC; Simms, Davidson, Lutski, 1898

The idea of armour plating military vehicles is undoubtedly as ancient as armour itself and so one hesitates to describe Hannibal's idea of protecting his **elephants** with sheets of metal in the 2nd century BC as an invention. It is an idea which the Mongolian emperor Kubilay took up again in the 13th century saying that it had never been done before.

Numerous historians attach great importance to a painting by Leonardo da Vinci of a circular blockhouse with a conical roof, which is set on wheels and equipped with a battery of cannons. This they are keen to see as the first **tank**. In reality, these kinds of fantasies, which probably only existed on paper, are common in drawings by Renaissance engineers, from the **armed chariots** of Marianus Jacobus, known as Il Taccola, and Francesco di Georgio Martini's **covered vehicles** in the 15th and 16th centuries. Vinci's blockhouse, however, from its sheer size with its armour plating and battery of cannons would have been such a weight that it would have been difficult to move, unless it had half a dozen, or even a dozen horses to pull it from the inside. If it was ever built, and we have no accounts of this, its effectiveness cannot have been convincing, for up until the 19th century and whilst **maritime armour plating** was being developed, no military leader seems ever to have referred to it. In actual fact, the armour plating of land vehicles was only made possible after the completion, in 1886, of an **internal combustion engine** by the German, Gottlieb Daimler. Only twelve years later, during the same year and independently, the Englishman, F. R. Simms, the American R. P. Davidson, and the Russian, W. Lutski, constructed an **armour-plated motor vehicle**, equipped with a **machine gun**. The **tank** itself was then being researched, but did not appear until the 20th century.

Artificial satellite

USSR, 1957

It is just as difficult to omit the artificial satellites from a census of inventions as to class them among actual inventions themselves. From the end of the 19th century to the beginning of the 20th century the perfect assimilation of Newton's laws had given the idea to numerous people interested in astronautics that a body equipped with a minimum speed and launched beyond the layers of the Earth's atmosphere, would continue to travel indefinitely round the Earth. This principle is explained in the two-part novel by Jules Verne, *De la Terre à la Lune* and *Autour de la Lune*, which was published in 1865, where the calculations remain correct to this day. All the same, sending a satellite into orbit required a sufficiently powerful **launcher** to reach what is called an **optimal escape velocity**, in fact 11.20 km/s, which would allow an altitude of approximately 200 km to be reached. The pioneers in this research, which was at once theoretical and practical, were essentially the Russian, Constantin Tsiolkovsky, the American Robert Goddard, and the German Hermann Oberth. During the 1930s the Germans were undeniably in the lead in terms of technical advances: the famous **V1** and **V2 Rockets**, which were in fact the precursors of space launchers. Part of the German team went to the USSR after World War II; another, led by Wernher von Braun, went to the United States. These two teams raced each other with their astronautical programmes. It was however the USSR

An artistic representation of a US Mariner satellite travelling close to the planet Mars.

that claimed the honour of inaugurating the Space Age with the launch of Sputnik 1 on 4 October 1957, a feat all the more remarkable because the USSR did not have the use of the computers which the Americans had developed.

The artificial satellite appears in this chapter on military technology because the first launchers were obviously created in a strongly political and military context; however the invention subsequently passed in to the civil domain for the purpose of **meteorological surveillance** and for relaying **telecommunications**.

Atomic bomb

Meitner, Frisch, Bohr, 1939; Neddermeyer, 1943

When the release of energy by the **fission of the atomic nucleus** was discovered by Hahn, Meitner, and Strassmann in 1936 (see p. 90) there was the undeniable implication that this energy could one day be exploited. At the time, however, the future methods of this exploitation, whether civil or military, could hardly be predicted. The Austrian,

Lise Meitner, who had to leave in the middle of her work with Hahn (she had been constrained by the 'racial' laws of the Third Reich), did not know the final results of the research until she was exiled in Sweden. She and her nephew, Otto Frisch, discussed the results obtained and had the premonition that they were dealing with the seeds of an atomic weapon. They feared that Germany would be the first to make such a weapon. A short while later, the British Government and the famous Dutch atomic physicist, Niels Bohr, were alerted to Lise Meitner's warnings and found them to be justified. In 1939, Bohr was sent as emissary to the United States with the task of putting forward Lise Meitner's theories, with the aim of persuading the Americans to undertake the work which would eventually lead to the production of an atomic bomb. So the invention of the atomic bomb can be attributed originally to Lise Meitner, Otto Frisch and Niels Bohr; at least, they were the first to conceive it.

There was still a long way to go before an explosive weapon based on the atomic fission could be realized; even the fundamental mechanisms which had first led to Hahn recording the release of energy, some 200 million electron volts, and the chemical process which was the source of it all were not known. It was only understood when Bohr, with the help of the American physicist, John Wheeler, repeated his experiments at the Princeton laboratory, bombarding uranium with **slow neutrons** as Hahn had done. It was established that it was the nucleus of **uranium-235** alone which had fissioned; the other isotope, **uranium-238** had only absorbed the neutrons. This meant that considerable progress had already been made; he demonstrated that an atomic weapon should be based on the fission of **uranium-235**. The principle of the energy-releasing explosion was thus established in 1939; if sufficient quantities of **uranium-235** were used, a **chain reaction** could be induced where the neutrons freed from the fission of the first nucleus would act upon other nuclei to split them, then these neutrons on yet more — the number of split neutrons growing with the reaction. Thus, inconceivable quantities of energy would be released, well beyond those recorded by Hahn.

Alarm grew among the physicists who were working for the Allies: at the same time Germany had no lack of brainpower capable of conceiving such a weapon. In fact, Hahn and Werner Heisenberg, among many others, had already thought of it. All that remained was to produce sufficient uranium-235 to carry out some experiments and then to devise the mechanical principle which would lead to the manufacture of a transportable weapon, that is, a bomb.

Normal uranium ore contains 0.7% uranium-235. So the Americans began by constructing a **centrifugation** plant for the ore, a process which was then replaced by that of **gaseous diffusion** across filters which would only allow molecules lighter than uranium-235 to pass through. The physicists then realized that the principle of gaseous diffusion constituted a very lengthy process, with the filtering having to be repeated thousands of times to obtain 90% pure uranium. Moreover, it turned out that the size of the plants required for this was immense. Then Edwin McMillan and Philip Abelson from the University of California discovered **neptunium** or **element 93**. They postulated, quite rightly, that it would decay to become element 94. Bohr and Wheeler also theorized that this element of mass 239, would split when bombarded by slow neutrons and so could be substituted for uranium-235.

Until then the US government had been sceptical about the possibilities of producing an atomic weapon from laboratory experiments; the means of physical and chemical processing were proving to be extremely expensive. Nevertheless, a letter from Albert Einstein to the President of the United States, Franklin D. Roosevelt, must have strengthened the determination of the responsible authorities. The **cyclotron** (particle accelerator) from Berkeley University in San Francisco was put to work to obtain workable quantities of element 94; in 1941 this element was established to be **plutonium-239** and its fissile possibilities were also recognized. It was only one year later that plutonium-239 was proved to produce more neutrons on splitting than uranium-235.

The path to the production of the atomic bomb was lined with considerable difficulties. The production possibilities of uranium-235 were reduced and as the research advanced the theoreticians raised their estimates of the 'critical' mass — that from which an explosion could be generated. Moreover, no one knew the chemistry of plutonium-239 and, when it was made, it was discovered to produce an unstable

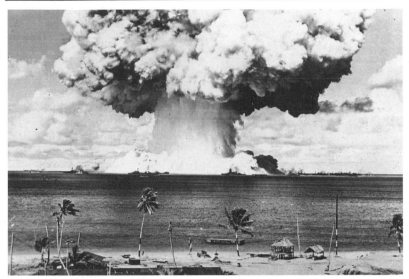

July 24th, 1946, Bikini atoll: the first underwater explosion of an A-Bomb, which sent a 600-m wide column of water 1500 m into the air.

isotope, plutonium-240, which fissions spontaneously and produces neutrons but in too small a number to generate a chain reaction. Finally, the explosion mechanism itself also posed a problem.

In fact, it was not enough simply to have a **critical mass** of uranium-235 or plutonium-239 to obtain a violent fission reaction to the **supercritical point**, that is, with the accompanying release of energy and chain reaction. These materials become capable of spontaneous neutron emission and are dangerous to the health, but their explosions do not take place within a specific time. Instead, they emit neutrons continuously (a fact that is ex-

The sense of urgency in pursuing the work on the atomic bomb by the allied forces was partly justified. If the German physicists, who were also trying to perfect a bomb of the same type, were acting rather slowly, the Japanese on the other hand were working diligently on the perfection of the A-Bomb. Paradoxically, it was the inconsistency of the military command in charge of such operations which was the greatest hindrance to the whole venture.

ploited in atomic power stations), but the critical mass only produces the expected explosion when there is minimal loss of neutrons, which in turn only occurs when the fissile material is enclosed in a reflective casing.

In order to bring about a critical mass which explodes at the precise moment, the technique adopted was one initiated by the American, Seth Neddermeyer, in 1943. It was crucial to the fabrication of the first A-Bomb: two critical half-masses were created and brought violently into contact using a classical explosive device; thus, the energy-producing fission reaction got under way. The device actually consisted of precipitating several sub-critical masses towards the centre of the reaction chamber. They combined with each other to form a mass which passed quickly to the critical point. In this way an **implosion** was obtained. The American physicist, Edward Teller, devised an improvement which consisted of placing these masses under pressure, so reducing the mass of material required.

Given that the production of uranium-235 had not yet reached the required mass, it was replaced by plutonium-239 and the first experimental explosion took place on 16 July 1945, under the code name 'Trinity'. Its magnitude surprised everyone: although the experts had predicted that the energy released by the explosion would be equivalent to that of an explosion of 1000 to 5000 tonnes of TNT, it was in fact equal to 20000 tonnes of TNT. The first atomic bomb, 'Fat Boy', was made with uranium-235. It was never used for ex-perimental purposes, because the material of which it was made was in such short supply. When it was dropped on 6 August 1945 at 08.15 local time in Japan, it destroyed the town of Hiroshima. Japan declared her surrender only after the explosion of the second bomb, this time made of plutonium-239, which was dropped on Nagasaki on 9 August.

Beam weapon

Anon., United States, c.1975

A beam weapon is the name for a **strategic armament system** consisting of an **electron accelerator** designed to produce beams of rapid protons which will destroy solid targets, such as a satellite. This system seems to have been created around the year 1975 in the USSR and United States simultaneously, as shown on the one hand by contemporary experiments in issuing electron beams in space from the Soviet spacecrafts Cosmos, Soyuz and Salyut, and on the other hand by the U.S. Army Balistic System projects, entrusted to the firm Austin Research Associates.

The 'cannon' principle was to follow. To begin with, a diode at one end transforms an electric current with a power of 10^{11} W (one hundred thousand million watts) into electron beams, which are accelerated to a speed close to the speed of light; these beams move in spirals within a cage, along the lines of a magnetic field of force created by electric coils. Secondly, some **hydrogen atoms** (which consist of a nucleus or proton around which a single electron revolves) are injected into the cage 'downstream' from the accelerated electrons and the hydrogen is ionized — the peripheral electron is torn out of the hydrogen atom; only the proton remains. Thirdly, the protons are accelerated by the electron beams up to a distance of 4 m from the south of the 'cannon', so the proton beams follow their course to a destructive impact with a given target, at the rate of a blast of protons every 200 nanoseconds (thousand millionths of a second). The beam has a width of 1 cm.

Since the first experiments it has been a known fact that this type of weapon requires very intense and very powerful electric currents, and could only be brought into practice with the help of new types of capacitors (electrical components capable of storing electric charge) and generators capable of producing such currents in very short spaces of time. The chosen goal in the 1980s was to send out proton beams with a range of action equal to about 10 km, a useful distance from which to be able to destroy a missile in the atmosphere. At the same time, the making of capacitors and generators powerful enough but nonetheless small enough to be fitted to a satellite posed considerable problems; this was why a version of the beam weapon was tested by the U.S. Navy for use on ships, for which the freight of huge units of energy would pose many fewer problems. In the 1980s the weight of a unit capable of providing energy impulses of 100 MeV (million electron volts) for 10 nanoseconds, or 6 impulses per second for a distance of 500 m, was estimated to be 100 tonnes; so the general research philosophy favoured mini nuclear reactors as the energy source, as they are light enough to be put into orbit. It seems that this is the solution adopted by the Soviets who have been making such reactors since 1978 and are in general using them to equip their radar satellites.

Hydrogen bomb

Teller, Kourtchatov, 1952–53

The **nuclear hydrogen bomb**, usually known as the H-Bomb or **thermonuclear bomb** is one of the most terrible military inventions in history. The date assigned to it above is that of its actual completion; in fact, for the United States, the history of the invention began in 1942 when a conversation took place between Enrico Fermi, the famous Italian physicist, and the equally famous American physicist Edward Teller, who had taken part in the manufacture of the first atomic bomb (see p. 217). Fermi had suggested the possibility that **fission** such as that of the atomic bomb, might trigger a chain reaction comparable to those which are activated in the Sun. The theoretical study undertaken to verify their hypothesis proved a few months later that it was correct; the fuel to be used would be **deuterium** (an isotope of hydrogen). Subsequently it was concluded that a **deuterium–tritium** mixture would produce the best results because it would allow the critical temperature to be lowered, and the release of energy would be more powerful.

This principle differed fundamentally from that of fission, which amounted to a single explosion of a critical mass where the atoms were 'split'. The H-Bomb was, however, a matter of a fission being followed by **fusion** with a chain of reactions ensuing until the fuel was exhausted. Fusion consists of the combination of several light atoms into one single atom with a much larger release of energy than with fission.

All the same, the plan could only be set in motion with a reliable **detonator**, an atomic bomb, which was not tested until 1945. The programme of the H-Bomb followed after the war, but rather slowly; it was the experimental explosion of a Soviet A-Bomb on 29 August 1949 which rekindled a sense of urgency in the Americans. In 1950 a deuterium factory was opened in South Carolina. However, the theoretical research which was vital in the completion of the thermonuclear bomb came up against calculation difficulties: no **computer** existed which could cope with them. So Teller resolved to simplify the task considerably; besides, some of the reduced calculations had shown that much more **tritium** would be required than the initial plans had predicted. The research continued until the first thermonuclear bomb was put to the test on 1 November 1952, at the Pacific atoll of Eniwetok.

Although there was less available information on the whole subject, the Soviet thermonuclear bomb followed much the same course. Since 1942, when the first atom bomb had not yet been built, the Soviet physicists, like Fermi, had imagined the possibilities of fusion reactions. They worked away at it therefore, with all the more assidulity for having fewer means of making calculations than the Americans, with **electronics** having been a neglected area in Soviet ideology. Despite the insufficient safety conditions which were prevalent in the Soviet plants, and which caused the death of numerous engineers, the USSR succeeded in exploding a thermonuclear bomb on 12 August 1953, about eight months after the Americans. The man in

> The thermonuclear bomb's potential for annihilation, which was considerably greater than that of the atomic bomb, caused alarm among the superpowers who could not accept the risks of its use, neither for themselves nor with regard to international opinion.
> On 10 October 1963, the United States, the United Kingdom (which had produced its own thermonuclear bomb in 1957) and the USSR signed the trilateral agreement which banned this weapon. Despite the political efforts of the three superpowers then, the countries which had not signed the treaty pursued their own research and equipped themselves with thermonuclear weapons. On 17 June 1967, China held an experimental thermonuclear test, followed by France on 24 August 1968. Several other countries, notably India and Israel, have the necessary potential to make thermonuclear bombs.

charge of it all was the famous physicist Igor Kourtchatov.

The nuclear bomb or **N-Bomb** is derived from the H-Bomb and was devised in 1979 by the American physicist, Edward Teller, and the mathematician Stanislaw Marcin Ulam. It consists of a bomb of low power (between 0.5 and 5 ktonnes) which causes the death of all living organisms within a range of 800 to 1200 m. Its principle is simple: when the power of a nuclear explosion is reduced, the mechanical and thermal (blast and fire) effects diminish more quickly than the radiation.

Machine gun

Barnes, 1856; Ripley, 1860; Gatling, 1862; Maxim, 1883; Browning, 1885

Despite the invention of the revolving loader by the Englishman James Puckle in 1718 (see *Great Inventions Through History*), the machine gun was not developed until the middle of the 19th century.

In 1856, the American, Charles Emerson Barnes, took Puckle's principle and improved it considerably by adding a **crank** which automatically moved the obturator backwards and forwards. Real progress was made, for bursts of 60–80 shots per minute could be fired merely by turning the crank. This kind of machine gun, which was nicknamed the 'coffee grinder' when used in the American Civil War, was modified again by E. Ripley, a compatriot of Barnes, who introduced the **multiple barrel gun** with the same kind of crank, of which there were several other versions. But all these machine guns needed teams of two men — one server and an assistant who fed the bullets into the loader.

The critical stage in the evolution of machine guns was passed with the assistance of Richard Jordan Gatling, and **metal cartridges** (see *Great Inventions Through History*). The Gatling machine gun comprised six guns in a row, fed by a cylindrical drum loader situated above the weapon; a crank rotated a camshaft which itself, via a gearing system, rotated the loader; the weight of the cartridges themselves caused them to fall into the firing chamber, provided that the cam made the obturator move and armed the firing pin. The guns were loaded in this way from the top left, provided that at the bottom right the firing pin fired the shot. A complicated system, it nonetheless proved its efficacy for it could fire 400 shots per minute. As well as the server, an assistant was required to keep the loader constantly fed.

More progress was made in 1883 by the American, Hiram T. Maxim. It consisted in using the recoil from the reaction of the explosive gases, firstly from the black powder cartridges, then the smokeless cartridges. These gases made the gun and the obturator kick back; the gun is stopped by a block but the obturator continued backwards carrying the cartridge with it, which was then thrown out by a spring at the bottom of the breech. If the trigger had no pressure applied to it, the obturator stopped against a block; but if a finger remained on it, the obturator, also propelled by a spring, set off another cartridge in the gun and a new volley of shots was produced. The Maxim machine gun functioned therefore for as long as one kept a finger on the trigger and, with this improvement borrowed from the Bailey machine gun, it no longer needed an assistant because the cartridges streamed out of a **belt feed**. This

> The Gatling machine gun had five successive improvements in the form of the Nordenfeldt version (1879) which had volley fire; the Gardner of the same year for the marine gunners because it was light; the Lowell, also of the same year, which fired very fast for its time; the Wilder (1880) which was the first to be fed by a belt feed.

*A Spandau 08 15 **machine gun** and its group of German servers during World War I.*

was a **gas machine gun** and its efficiency, 600 shots per minute, impressed the army leaders around the year 1900. It was perfected in 1885 by the American, John M. Browning, who exploited the explosive gases to the full by collecting them in a cylinder where they activated a piston which in turn activated the obturator.

Since then, the machine gun has been improved many times, but not usually by new inventions, apart from the **synchronized firing with the aeroplane propeller** which permitted firing across the propeller (1915) and the lighter adaptation of it, the **machine gun** (1916). It was adopted in different versions by infantry, tanks and airforce, since the carriages, loaders and calibres were all variable.

Radio-controlled air-to-air missile

USAF and Hughes Aircraft, 1954

The first radio-controlled missile to be launched from an aeroplane was made jointly by the US Air Force and the firm Hughes Aircraft. It was destined for the Northrop fighter plane F89 Scorpion and had the sign GAR on it. This kind of missile is based on the principle of **infra-red (or heat) detection** which, with the help of an in-built computer, the missile continually compares its original trajectory with the infrared point provided by the heat from the motor of the target aircraft. This equipment, the **directional platform**, uses **active infrared**, since it employs an infrared beam which lights up the target; the retification of the flight path is effected by the superposition of the beam and the infrared radiation of the target. A parabolic mirror concentrates the reflected rays on to a **bolometer** (detector) which confirms the actual location of the target.

Swing-wing aeroplane
Grumman, 1954

The US aeronautical firm Grumman was the first (in 1954) to devise and build an aeroplane which could change the angle between the wings and the fuselage from the T-winged subsonic position to the supersonic delta-winged position: the aircraft was the XF 10 Jaguar of which only two models were built. The advantage of this mechanism, the **swing-wing**, was to allow the supersonic jets a **short take-off**. It would also allow a supersonic bomber on a military mission to fly very low with the least fuel consumption.

Though it was studied by numerous aeronautical designers during the ensuing quarter-century, interest in the swing-wing

One of the last swing-wing aeroplanes still in use, the Mig23S, during a demonstration at Kuopio in Finland on 2 August 1978. With the wings outstretched this Soviet aircraft could fly at Mach 1.1 at low altitude, and with the wings folded back could reach its maximum speed of Mach 2.3.

gradually disappeared, due to the growing specialization in aircraft. By the end of the 1980s no more than three prototypes of swing-wing aeroplanes remained — the Mig23 and the strategic bomber 'Black Jack', both of Soviet manufacture, and the American strategic bomber B1, whose fate was uncertain.

Synchronized propellor and machine gun
Fokker, 1915

The positioning of machine guns on the first military aircraft had to be either above the rotation zone of the propellor, that is, on the upper wing, or on the side of the aircraft, which affected the precision of firing.

It was Anthony Herman Gerhard Fokker, a Dutchman, who, whilst working for the Germans during World War I, succeeded in exactly synchronizing the rhythm of the machine gun with the number of revolutions per minute of the propellor. This was achieved with the help of his own **differential** system which allowed the machine gun to be installed directly within line of the pilot's vision. When the first aircraft, the Fokker Eindecker, a monoplane, was equipped in this way, it caused devastation amongst the allied ranks. Nine months later Georges Constantinesco, a Frenchman of Romanian origin, reconstructed Fokker's invention and thereby gave the Allies the means to reply in kind.

Index

Note: Entries in **bold type** refer to articles headed with the name shown.

INDEX